MW00616548

THE RISING

THE RISING

The Twenty-Year Battle to Rebuild the World Trade Center

LARRY SILVERSTEIN

ALFRED A. KNOPF · NEW YORK · 2024

THIS IS A BORZOI BOOK PUBLISHED BY ALFRED A. KNOPF

Copyright © 2024 by Larry A. Silverstein

www.aaknopf.com

Knopf, Borzoi Books, and the colophon are registered trademarks
of Penguin Random House LLC.

Library of Congress Cataloging-in-Publication Data
Names: Silverstein, Larry, 1931– author.
Title: The rising: the twenty-year battle to rebuild the World Trade Center / Larry Silverstein.
Description: First United States edition. | New York : Alfred A. Knopf, 2023. | Includes index.
Identifiers: LCCN 2024002646 | ISBN 9780525658962 (hardcover) |
ISBN 9780525658979 (ebook)
Subjects: LCSH: Silverstein, Larry, 1931– | World Trade Center Site (New York, N.Y.) |
Memorials—New York (State)—New York. | City planning—New York (State)—New York. |
New York (N.Y.)—Buildings, structures, etc.
Classification: LCC F128.8.W85 S55 2023 | DDC 974.7/1044—dc23/eng/20240524
LC record available at https://lccn.loc.gov/2024002646

Jacket photograph by Joe Woolhead
Jacket design by Chip Kidd
Book design by Jo Anne Metsch

Manufactured in the United States of America

First Edition

To my wife, Klara, the joy of my life.
Sixty-eight years and still going strong!

DISCLAIMER

This is a work of nonfiction in which the author has reconstructed dialogue to the best of his recollection.

THE RISING

PROLOGUE

THE DOCTOR DIDN'T SAVE me; the appointment did. Or, more precisely, my wife's insistence that I keep it. If I hadn't listened to Klara on that fateful morning of September 11, 2001—the day the world changed for me and so many others—things might have turned out very differently.

When I woke up that bright late-summer Tuesday, my intention was to head downtown from our apartment on Park Avenue and have my usual eight-thirty breakfast at Windows on the World. This was the restaurant on the 107th floor of the North Tower of the World Trade Center in Lower Manhattan. It'd been my routine for the past two months to use these breakfasts to meet with the buildings' major tenants. Now, you might think that would take years; the towers, after all, contained 10 million square feet of office space, enough to fill 174 football fields laid out goalpost to goalpost. However, only forty companies accounted for about 70 percent of the occupants. And so I had already made my way pretty far down the list and had gotten a good sense of how many of the primary renters felt about working in the complex. It had been a valuable education. And while

Windows, even at breakfast, was often fully booked, I could always be sure of getting a good table, one looking north up the blue ribbon of the Hudson River with the greatest city in the world, vibrant and splendid, spread out before me.

You see, I owned the Twin Towers. "Owned" is actually a bit of shorthand for a more complicated contractual reality. On July 24, 2001, just fifty days earlier, after signing a 1,160-page agreement with the Port Authority, the joint venture of the states of New York and New Jersey that controlled the sixteen acres of land and the buildings that constituted the World Trade Center complex, New York governor George Pataki had handed me a set of oversized cardboard "keys" to the Twin Towers. And while the gigantic keys were a bit of vaudeville humor from the usually dour governor, I did in fact now have a ninety-nine-year lease on what were at the time the two largest buildings in the world. And I had previously handed over a $3.2 billion check for this leasehold. My monthly rent bill was $10 million.

The weekday breakfasts, then, were pretty important to me. One of the first rules I had learned in what had been up to that point nearly a half century in the real estate business is that it's crucial to keep your tenants happy. You should get to know their needs, learn how you can help them. It's a process; both landlord and tenant must get to know each other. You just don't come in one day and announce that you're raising the rent.

However, that morning another of the cardinal rules I had learned after a lifetime of experience took precedence: when my wife insists, my answer is always a succinct "Yes, dear." It's an insight that has kept me very happily married for the past sixty-eight years.

I had previously scheduled a nine a.m. appointment with my dermatologist, whose office up on Sixtieth and Madison was just a short walk from our apartment. It was my running joke that I was her annuity; she was always finding something troublesome that needed to be removed. But I guess that was the price I had to pay for being fair-skinned and for my favorite hobby.

Back in 1960, after Sharon, our first child, had been born (Roger would arrive three and a half years later; Lisa in 1966), we were spending the summer away from our stuffy apartment in Washington Heights at a rented house in Stamford, Connecticut, and on a whim we had bought a boat for $2,500. We called it *The Last Penny* because that pretty much summed up our situation after our impetuous purchase. But Klara and I discovered we loved being out on the water, and in the winters we diligently spent a good deal of time taking maritime instructional courses. As our family grew and my business prospered, and even after Lisa was born and we moved to a spacious five-bedroom house in White Plains, we continued to spend vacations on the water—and over time upgraded to bigger and bigger boats. Our newest, the *Silver Shalis*—named after our daughters, Sharon and Lisa—is a 180-foot, four-deck custom-built beauty with a gleaming cerulean-blue hull. It's a very comfortable floating home with all the amenities you could ask for, including a gym, a small reverse-current pool, and even an elevator, as well as a full-time crew and a state-of-the art navigation system that could guide an ocean liner through a storm. We've happily cruised all over the world, and during these voyages I've continued to spend too much time exposed to the glaring sun. The frequent visits to the dermatologist were the annoying price I had to pay for my blissful holidays on the water.

But on the morning of September 11, as I was getting dressed, I abruptly decided to cancel my doctor's appointment. There was just too much work to do downtown at the Trade Center. As the new landlord, I should be meeting with tenants at breakfast, not spending time at the dermatologist. I told Klara that I'd reschedule.

"You canceled last time," she reminded me forcefully. "This is your health. You can't cancel again."

I could see she was upset. Very upset.

And so I didn't go down to Windows on the World that morning. I was not seated at my usual table on the 107th floor of the North Tower when the first plane crashed into the building at 8:46.

Instead, I was preparing to leave for the dermatologist when the phone rang. Klara answered it.

Larry and Klara Silverstein

"Is Mr. Silverstein okay?" the captain of our boat asked anxiously. "Yes," said Klara. "Why would you ask?"

The boat was docked at Chelsea Piers on Twenty-Third Street. Standing on the deck, the captain could see the plume of dark smoke escaping in a thick, ominous cloud from Tower One.

But he didn't try to explain. Instead, he tersely instructed, "Turn on your television set."

WHAT CAN I SAY about the rest of that day? The precise details are buried deep in my mind; even decades later they are apparently too painful to relive fully. What I do recall is a series of opaque images, the events blurred like photographs that were never fully developed. My memory has always been one of my prides, and I would often boast about it. But when I try to look back on that grim day, I'm overwhelmed. I can recall only moments, intimations of what I'd lived through.

There I was still in the apartment, Klara and I watching the television in stunned silence when the second plane abruptly turned north and slammed into the South Tower. *How can this be happening?* my mind screamed.

And what about my children? Both Roger and Lisa worked in the Trade Center complex. And the people who worked for me? We had temporary offices on the eighty-ninth floor; I hoped they hadn't arrived before nine.

When I tried to call their cells, the office—nothing. All the phones were down in Lower Manhattan and it would remain that exasperating way all day. I wanted, *I needed* to know that my children, my employees, my tenants were alive—and I could not reach anyone.

Then the towers fell. They started to crumble—and just like that they were gone. One moment they were there, and then they weren't. I was absolutely staggered. I couldn't believe what I was seeing with my own eyes. Yet in the next instant my bewilderment gave way to prayers. Where were my children? My employees?

Suddenly I realized I needed to get to our office at 521 Fifth Avenue, on Forty-Fourth Street. It became very important, the only thing that mattered. I wanted to get some sort of control over the rushing surge of events and I unreasonably decided that if I was in the office, seated at my desk, order would be restored. The world I once knew would be reinstated.

When I arrived, Ann Tobin, who had been my assistant for thirty-seven years, was already there. She knew better than to try to speak. All she could do was shake her head mournfully.

Throughout the day, our conference room began to fill with people. They were the wives and husbands of our employees. It was a community of strangers suddenly linked by common prayers, by wishful hopes.

My children, thank God, made it to the Fifth Avenue office. Both Roger and Lisa had stories to tell, tales of fortuitous escapes. Their arrivals created a surge of hope in the anxious relatives of the missing who were crowded into the conference room.

Then at around three p.m. Bill Dacunto, the man who was second in command of all our buildings, arrived. He was completely covered in ash, a gray, ghostlike presence. I couldn't recognize him at first.

"Bill?" I finally ventured.

He nodded mutely.

"How did you get here?"

"Walked."

"From downtown?" I asked incredulously. It must have been five, perhaps six miles.

"It was the only way," he explained with resignation.

I led Bill into the conference room so he could share what he knew, what he had seen with those who had been waiting for hours and hours for some word from their loved ones. It was a mistake. When they saw him—weary, saddened, blanketed in the thick white soot that was all that remained of the buildings where their husbands and wives worked—they started to realize that all would not be set right, that the day would not have a miraculous ending.

We lost four of our people that day, parents to six children. All in all, 2,753 people were murdered as the Twin Towers came crashing down.

IT WAS JUST AFTER seven the next morning when Governor Pataki called me at home. I was already awake; sleep had not been easy that long night.

"Larry," he began, "we need to talk. What do you think we should do?"

I answered at once. "We need to rebuild. There's not a doubt in my mind." I had not thought this through before he telephoned, but I was nevertheless utterly certain. We—Americans, New Yorkers— could not allow the terrorists to triumph.

"Okay," the governor agreed. "We need to start thinking about this."

"Absolutely," I reiterated. "We have to rebuild, or else they will have won. I am totally committed to getting this done," I pledged. The words were spoken with the binding force of a solemn vow, but little did I know what this commitment would demand. I had no idea, nor in truth could I have imagined, that for the next two decades I would be caught up in an arduous and all too often frustrating process, a series of head-to-head battles fueled by politics, power, and money—with nothing less than the future of New York City as the economic capital of the world at stake. I would witness colossal government ineptitude, the promiscuous spending of public money as years blithely passed and the construction overruns totaled billions of dollars. I would go to war against intransigent insurance companies to force them simply to honor the contracts they'd signed, and the millions of dollars in premium checks they'd cashed. I would see my name and reputation besmirched in the press as cynical politicians scrambled to deflect the blame for their mistakes and delays.

I would also have the joy of working with visionary architects, inventive engineers, brigades of determined construction workers, my own wonderful team, and even a few resourceful and responsible public servants. And in the long and arduous process, the site of the terrorist attacks as well as Lower Manhattan itself would be revitalized. Four majestic office towers used by over fifty thousand workers, along with the evocative September 11 Memorial and Museum, an architectural masterpiece of a transit hub and busy shopping mall where 250,000 commuters pass through each day, and a new performing arts center would be built on the ruins. And two more towers (one will include floors of residential apartments) are in the works. The surrounding neighborhood, too, has come alive with families, schools, stores, restaurants, and hotels. A downtown community once reeling from the terrorist attacks is now prospering, buzzing with a half-million people.

The account that follows is the true story of the consequences of my spontaneous yet heartfelt pledge made early on the morning of September 12. It is a candid and very personal chronicle of my more than twenty-year mission to rebuild the forbidding, charred pit— Ground Zero—that the World Trade Center complex had become, and bring Lower Manhattan back to life.

The World Trade Center site after the 9/11 terrorist attacks

PART I

THE KEYS TO
THE KINGDOM

ONE

LOOK UP THE WORD "keystone" in the dictionary and you'll find that it has two meanings. It's a shape: a wedged, trapezoidal form. Or it can refer to the final piece that holds everything together: the tie that binds. As the years passed, the "keystone site," as it was called, would carry both meanings for me. Originally, it simply referred to an irregularly shaped downtown building site. But in time it would be the glue that cemented my professional dreams, the impetus for larger ambitions.

Let me explain. Back in the late 1970s, it was this oddly whittled footprint of land known as the keystone site—a 53,000-square-foot downtown parcel—that had first led to my involvement at the World Trade Center. It was the one remaining undeveloped site in what was then a six-building complex, and it was owned by the Port Authority.

ESTABLISHED BACK IN 1921 by a compact between the states of New York and New Jersey that was authorized by Congress, the Port had grown over the decades to become an incredibly powerful and well-financed quasi-governmental agency. The broad language of the original agreement had enabled the agency to wind up controlling much of the region's transportation infrastructure. It builds toll bridges and tunnels between the two states, operates Port Newark, which handles the largest volume of shipping on the Eastern Seaboard, runs the Port Authority Bus Terminal in Manhattan and the PATH transit rail line between the two states, and administers the region's three major commercial airports. In addition, it had entered the business of commercial real estate development when it had erected the original World Trade Center towers. In fact, it had the extraordinary legal power to condemn any private land it wanted to build on. It even had its own two-thousand-plus-member Port Authority Police Department.

In the process it had become fantastically rich. Able to issue its own bonds backed by the seemingly endless stream of revenue from its tolls on bridges, ports, and airports, the agency was financially independent of either state; its bonds were paid off by the self-replenishing stream of tolls and fees, not by taxes. Its annual revenue today is a colossal $4.8 billion.

It was also swaggeringly powerful. Under law, the New York mayor and other city officials have no authority to intervene in the agency's activities. It's ostensibly controlled by the (often adversarial) governors of New York and New Jersey. They each appoint six of the agency's Board of Commissioners, and, by tacit agreement, the governor of New York selects the executive director to supervise the day-to-day operations. But governors and their appointees come and go. The real power at the agency resides in the vast administrative bureaucracy that year after year keeps the bridges, ports, roads, and airports running.

It had been the Port Authority that, back in 1965, had brandished its power of eminent domain to buy out the dozens of small shops selling used radios and war surplus electronics that filled a busy downtown Manhattan neighborhood called Radio Row. Bound by Vesey, Church, Liberty, and West Streets, the area's stores were demolished, streets disappeared, and on the now cleared sixteen acres the Port Authority went ahead with what had started out as a $400 million plan (about $2.5 billion in today's dollars) to build the world's first international trade center—a commercial complex for multinational businesses. It was an idea that had been percolating since 1946, when the New York State Legislature had passed a bill authorizing Governor Thomas E. Dewey to move forward with the project.

But it was not until thirty years later, in 1973, that the two looming monoliths on the Manhattan skyline, the North and South Towers of the World Trade Center, were completed. They were gigantic, 1,368 and 1,362 feet high, and had cost more than twice the original estimate, a staggering $900 million by the time they were done. To my eye they resembled, as one architectural critic dismissively put it, "glass-and-metal filing cabinets." But the buildings immediately became New York icons, symbols like the Empire State and the Chrysler Buildings of the city's irrepressible energy and innovativeness. By 1975, the Port Authority had completed four other smaller, yet nevertheless still sizable, buildings on the site. The complex had its own zip code, housed over four hundred companies, and an estimated fifty thousand people worked in its buildings. It was bigger than many cities.

But the Port Authority still had not finished building out its master plan. There was one site remaining for an office building— the keystone site. The Port's architects had already designed a 1-million-square-foot building, forty floors high, to be erected above the load-sharing foundation of the existing Con Edison electrical substation that provided power for the entire Trade Center as well as for most of Lower Manhattan, including Wall Street. Yet, after having gone through the difficult and unexpectedly expensive process of constructing the first six buildings in the complex, New York governor Hugh Carey decided the time had come for the Port Authority to get out of the construction business.

Responding to Carey's wish, the Port placed a small ad in the trade papers announcing that it would accept sealed bids to develop

The original World Trade Center

the site. It specified that the chosen developer would erect the pre-designed building, lease it out to tenants, and in return pay ground rent to the Port for the privilege.

Silverstein Properties was certainly not the biggest developer in the city. In fact, when measured against many national and international real estate development firms, we were rather small. But my staff and I looked at the numbers, and we decided there was a real opportunity here: the chance to erect a moneymaking building in an area of the city that would only grow more attractive to tenants. A million-square-foot building—well, I found the prospect irresistible.

And there was also something else motivating me: it would be a return to my roots. Nearly a quarter of a century earlier I had started out leasing—and then had moved on to buying and renovating—office buildings with my father in Lower Manhattan.

MY FATHER, HARRY, HAD wanted to be a concert pianist. He quickly realized, however, that he couldn't make a living as a musician. Instead, he went into the real estate business and at night he'd put my sister and me to bed in our apartment up on 183rd Street and Fort Washington Avenue by playing Chopin. My mother, Etta, was his secretary, bookkeeper, and, since she spoke fluent Yiddish, Russian, and German, as well as English, the translator for his many interactions with immigrant tenants. When my parents weren't talking about the office, they would be talking about music, or at least that's how it seemed to me.

I, too, had inherited their love for music and played the piano like my father. I attended New York's High School of Music and Art, making the long subway ride from Upper Manhattan to Midtown each day. But I had no delusions that I would make my way through life as a musician; I just wasn't as talented as many of my fellow students. So when I graduated, I went to NYU, the uptown campus in the Bronx. And from there on to Brooklyn Law School. Still, I never really wanted to be a lawyer, either. I passed the bar, but I never practiced. The value of a legal education, as I saw it, was to learn how to think and how to reason my way through complex real estate transactions. And to my mind one of the most positive attributes of Brooklyn Law School was that the classes ended at noon. That allowed me to hurry to the subway and then spend the rest of the day at Harry G. Silverstein & Sons, my father's company at 366 Broadway in Lower Manhattan.

We—along with my sister's husband—were in the business of brokering office leases for the dozens of small concerns congregated between Chambers Street and Astor Place that sold fabric remnants.

These days, this downtown area of New York, with its cast-iron, brick, and stone storefronts and loft buildings, has been inventively revitalized into the bustling commercial and residential neighbor-hoods known as SoHo and NoHo. But at the tail end of the 1950s, when Silverstein & Sons was trying to make its way in this part of the city, the lofts and office buildings in the area were tenanted largely by businesses tied to the city's bustling Garment District, which was located farther uptown between West Thirty-Fifth and Fortieth Streets.

These downtown satellite enterprises were pretty hardscrabble concerns, working with literally the scraps—or, to use the more tony phrase preferred by those in the trade, remnants—of the uptown clothing manufacturers. Silk, ribbon, wool, and swatches of fabrics were sold from these lofts and storefronts. It was pretty much an immigrant's line of work—only the immigrants kept changing as

Larry Silverstein and his father, Harry

upheavals in Europe and elsewhere in the world brought their waves of newcomers to the city. At the start of the century, the workforce was predominantly Jewish immigrants from eastern Europe. Then, the Irish and Italians came. And as my father and I set up shop downtown, the loft warehouses and storefront display rooms were receiving an influx of Hispanic laborers, men and women newly arrived from Puerto Rico and Mexico.

It was this constant influx of immigrants that gave this urban neighborhood of stolid early nineteenth-century cast-iron buildings its cosmopolitan vibrancy. I'd walk the streets and I'd hear people chattering away in a babel of languages, on one corner Yiddish, and on the next Spanish. It was a microcosm of a city in transition, a New York where one clan of newcomers after another was rolling up their sleeves and getting down to the hard work of earning a living to support their families in the city that had become their new home. It was an animated, striving, exciting place to be.

But for me, I quickly discovered, brokering leases in the neighborhood was a terrible way to make a living. You would put in four or five months negotiating a lease, and the best you could hope for was a $500 or $600 commission. No one could live on that.

Or, as was also on my fretful mind that busy winter in 1955, afford to buy an engagement ring.

KLARA WAS MY BOSS when I first met her back in 1951, and so, as I like to tease, our relationship hasn't changed much over the ensuing seventy years. We met at a summer camp in Hancock, New York, run by the Brandeis Institute. Then a senior at NYU, I paid for my stay in the country by working in the kitchen; there was no other way I could've afforded the tuition. Klara Apat, a junior at Hunter College, was in charge. And she was a stern taskmaster. She made sure everything was spotless. By the end of the summer two things were clear: I had developed a bad case of dishpan hands, and I knew I didn't want to see her again.

Then that winter I received a call from her parents. They were hosting a surprise birthday party for Klara and they had invited many of the kids from camp. I was reluctant to attend; they lived near the Brooklyn Botanic Garden and it would be a long trip by subway. But her mother said she'd be cooking a big meal, and I figured it would be foolish to turn down a free dinner, even if the price was a nearly two-hour journey.

The meal turned out to be a feast, and all homemade. I found myself wondering if Klara could cook like her mother. And now that she wasn't bossing me around as she'd done at camp, I found that I liked her smile. In fact, I liked everything about her.

We started dating; I was twenty-one, Klara two years younger. I had no money, so we spent a lot of time riding back and forth on the Staten Island Ferry. It was a cheap date, but it allowed us to talk, to get to know each other. And the starry nights, the skyline of Lower Manhattan as our backdrop—well, it was very romantic. I always had a lovely time with Klara.

Then at the end of one date after we'd been seeing each other for a while, I offhandedly asked Klara what she'd like to do next weekend. She announced that she wouldn't be in New York. She was going away. "For how long?" I asked anxiously. "The entire summer," she answered.

During that long summer I decided I couldn't let Klara ever walk out of my life again. When she returned, we agreed that we would date no one else. I gave Klara the pin I had won when I was elected into the NYU honor society. I had wanted to give her an engagement ring, only I couldn't afford to buy one.

IT WAS A BAD business deal that helped me realize how I might make the money I needed to buy Klara a ring. I had been working for months putting together a lease for a small fabric firm that was moving into a new space in a downtown building. When it came time for the lease to close, I anticipated pocketing a $600 commission. It would be a paltry return for a lot of diligent work, but I was excited; after all, I had plans for the money. Except at the closing, another person showed up. He had been the broker for the building's lease and he wanted his commission, too. The building's owner, with a wisdom that he considered Solomonic, decided to cut the commission in half. He would be paying me only $300. The other $300 would go to the broker.

That came as a shock. But I also could do the math, and I quickly realized there was more money to be made as a broker selling buildings rather than leasing them.

My first sale was 620 Broadway. I earned a $1,000 commission. I used it to buy an engagement ring. Klara still wears it to this day; she never wanted to get anything more opulent.

After we married in 1956, it was Klara who supported us. I was scrambling from deal to deal, but Klara earned a guaranteed $3,200 a year as a teacher in the New York public school system. It was her salary that enabled us to rent an apartment for $110 a month on Chittenden Avenue in Washington Heights and buy a $2,000 Chevy Bel Air. When she got her master's degree and her salary was increased to $3,600, I felt like I had married a rich woman.

But we were also talking seriously about starting a family, and that meant Klara would need to stop teaching. I knew I had better find a way to make some money, too.

AS FATE WOULD HAVE it, just about the time I was looking to earn some money, I had heard what struck me as an incredible story. Harry Helmsley, a broker, and Lawrence Wien, an attorney, had managed to buy the Empire State Building, the tallest building in the city, *without putting up any money of their own.* They got 3,300 people to invest $10,000 each and formed a limited partnership to purchase the skyscraper. And if they could buy a building with other people's money, maybe, I thought, I could do it, too. I made up my mind to try.

I began going around Lower Manhattan, knocking on doors, asking owners if they'd want to sell their buildings. It took me nearly six months of trying, but I finally found a dismal building at 220 East Twenty-Third Street. It was a dark and derelict structure, but it had fourteen stories, space for plenty of tenants, and, I believed, lots of potential.

"Would you be interested in selling?" I had asked the owner.

"I'd like to try," he said. "Pay me fifteen thousand dollars up front against the purchase price, and I'll give you the right to sell the place." But he warned me: he wouldn't settle for less than $600,000 as the sales price. The way he'd announced the number, though, made it clear he thought it was an amount no reasonable person would ever pay. But I wasn't so sure. To my mind, it seemed like a very attractive price, maybe even a bargain.

I was enthusiastic and rushed back to the office. "Dad, I got a chance to broker the rights to sell a building. But why should we settle for that? Who wants to be a broker when we can be an owner? Let's see if we can buy it!"

The problem with that, I soon discovered, was that no bank was willing to lend us the initial $15,000 seed money necessary to set this plan into motion. And why should they? We had zero collateral to offer lenders. Not a penny.

I had pretty much given up on this pipe dream when my father and I were walking back to the office after a lunch at Ratner's, a kosher dairy restaurant on the Lower East Side (I can still taste their onion rolls), and we passed a bank at 85 Delancey Street. I thought we had approached every bank in Lower Manhattan, but the Public National Bank was one we had apparently missed. Might as well give it a try, too, we decided on a whim.

We introduced ourselves to Phil Green, the manager, told him about the building we wanted to buy, and how we wanted to borrow $15,000.

"Do you have any collateral?" he asked.

"If we had, we wouldn't be here," I replied truthfully.

The bank manager laughed and then reached into a drawer and handed me a legal form.

"What is this?" I asked with considerable skepticism.

"A promissory note."

I was incredulous. "Are you going to give us the money?"

"Isn't that what you want?" he replied.

"Dad, sign this piece of paper. Quick!"

He did. And after we had the money and paid the owner $15,000, we had secured the right to purchase the thirteen-story building for $600,000. Only now we needed a large mortgage.

This time when we approached banks, however, we had the building to use as collateral; that is, it would be our collateral if they granted us the mortgage. We went hunting, and, finally, a savings bank on Twenty-Third Street agreed to give us a $350,000 first mortgage.

The deal would not have succeeded without Klara's help. To get the financing we needed, extremely precise amortization tables were required. Klara spent the entire summer putting them together, and when she was done the bank was very impressed. But that left another $250,000 we still had to raise.

It took us a couple of months, but we finally found a bank that was willing to send an appraiser to look at the building to see if it was worth an additional quarter of a million dollars.

I was in the office when I got a call from the appraiser, Leon Spear. "Mr. Silverstein, I just went over the building and I'm willing to give you a $350,000 mortgage—if you'll sell the place."

We had needed a quarter of a million dollars, but now out of the blue he was offering us $350,000—and that would allow us to make a $100,000 profit if we sold him the building immediately after becoming the owners. It was more money than I could imagine. A fortune. And we wouldn't have to do anything except sign over the title.

"Dad," I rejoiced, "we have an offer that will make us a hundred thousand dollars!"

My father was less enthusiastic. "If it's good enough for him, I think we should keep it. Turn the building into something."

I persisted. "Dad, this is a hundred thousand dollars. How can we turn it down?"

"If it were me alone," my father explained, "I'd take the money and that'd be that. But you're young. Here's your chance to start something and make some real money."

My father's words made sense. I, too, realized this was a genuine opportunity—but only if we could find some way of raising the $250,000 in equity we needed.

"Could you figure out the details?" he wondered.

I quickly got a copy of the offering brochure for the Empire State Building that had prompted my initial foray into the real estate business and studied it with a renewed interest. It became my guide for financing the purchase of East Twenty-Third Street as a limited partnership.

Sure, our building wasn't anything as grand or as expensive as the Empire State Building. But the same financial principles that had worked so successfully for Wien and Helmsley now worked for us, too. Instead of selling our contract, we found twenty-five investors, people with whom my father had done leasing business in the past, and got them to put up $10,000 each as limited partners in the ownership of the building. These transactions allowed us to raise the entire $250,000 we needed to complete the purchase.

It was 1957, and at twenty-six I had been married for almost two years, and I now owned my first building. I had never been so excited, so happy. It was an incredible feeling.

And I was just getting started. New plans for our purchase were taking shape in my mind. "Dad, we need to fix the place up," I urged, and he agreed. We sanded floors, cleaned the windows, painted the walls a bright white, got the rickety elevator working in a dependable fashion, and did up the gloomy lobby so the entrance was more inviting.

When we were done, we'd transformed the building into some-

Larry and Harry Silverstein bought their first building at
220 East Twenty-Third Street in 1957 for $600,000.

thing bright and shiny. An address where tenants would want to
locate—and where we could charge a higher rent than we'd previously imagined.

This new income allowed us to refinance the existing mortgage—
for considerably more than $350,000. And now we were able to return

some of our investors' initial investment capital as well as give them regular distributions. It was terrific: we were making money and so were our investors. And in the happy process I discovered one of the inflexible rules of the real estate business: a satisfied investor is someone who'll not only want to do business with you again, but will get his friends to put up money, too.

I was beginning to see that this was a financing strategy that had real potential. Our next purchase was on East Forty-Seventh Street, for $1.5 million. It took a bit of time and persuasion—it was more than twice as expensive as the purchase of our Twenty-Third Street building—but the limited partnership model worked once again. We were able to get our previous investors to hand over the money we needed to close the deal. It was a big step forward for us.

And it encouraged us to aim higher. Our next purchase was on East Sixty-Third Street, for $2.5 million. And now Harry G. Silverstein & Sons was on its way. By 1966, the year my father died, we owned nine or ten small buildings.

At this point people in the business were noticing what we were doing. I got a call from the bank on Delancey Street, the one we'd walked into by chance and that had lent us the $15,000 we needed to buy the right to purchase the Twenty-Third Street building. They announced that they'd like to be our bankers. Come to us when you find another property that needs financing, they advised. And we did. We no longer needed the small investors. The banks were willing to fund us. It was terrific.

By now Klara and I had three children. We decided to move out of Washington Heights and settle in the suburbs. I bought the five-bedroom home in White Plains, New York, for $71,000.

It was a wonderful, comfortable home. The house was always full of kids, our three and their friends. Klara made sure that there was a jar of fresh-baked cookies and a refrigerator full of cartons of milk; it was the neighborhood hangout. And a home for a happy family.

And all the while my real estate business kept growing. But one ambition was still beyond my grasp: I hadn't built a building.

IN THE EARLY 1970S, I got my chance, though for a while, I regretted it. I had purchased a site near the Garden State Parkway in New Jersey, a busy commercial area near the Bell Telephone Laboratories offices. The parcel was big enough to accommodate a 300,000-square-foot building, and so that was what I wanted to build.

But I didn't dare say that. It seemed too large, too ambitious an undertaking. Instead, I put up a sign on the property modestly offering "100,000 square feet—expandable." I didn't want to scare people away.

But it apparently did. Nobody made any realistic inquiries.

I was growing anxious, yet I nevertheless decided to move forward. I've never been afraid to gamble if I was convinced the ultimate prospects would be favorable. You need that sort of nerve if you want to prosper in the real estate business. And so without having even a single tenant for the prospective building, I started clearing the site for construction.

It wasn't long after a building shed went up on the property that my phone rang. "How large a space can you build?" the prospective tenant asked. "Three hundred thousand square feet," I announced with some trepidation. The tenant wound up taking the entire space for twenty years. With this lease agreement in my pocket (banks, I was learning, need to know you can rent the space before they'll write a big check), I was able to get financing. I could now build what back then seemed like a gigantic office building.

Yet, as this building got underway, it was also a time of some sadness. In 1977, Bernard Mendik, my partner since we'd both started working for my father over twenty years earlier, and my sister, Annette, had decided to divorce. Their split made it impossible for me to remain in business with Bernie; sibling loyalty came first. We reached an agreement to divide our holdings, which by now included five buildings on Fifth Avenue, one on Wall Street, and a shopping center in Stamford, Connecticut, among several other choice properties. But Bernie and I had also reached a divergence in our business strategies. He wanted to buy buildings. I wanted to build them. The experience in New Jersey had whet my appetite.

AND SO IT WAS with a good deal of eagerness that I submitted my bid to construct a 1-million-square-foot building on the last remaining piece of land in the World Trade Center complex. I didn't think our company would have much of a chance against the bigger, more experienced competitors, but we won the keystone site.

T W O

I THOUGHT IT WOULD BE a fairly simple business to negotiate the
lease with the Port Authority; after all, they'd approved our bid.
And then it wouldn't take long to find a tenant, which would
allow me to get the financing needed to begin construction. As deals
go, this would be a pretty quick and uncomplicated transaction.

I couldn't have been more mistaken.

AT FIRST, THE ANNOUNCEMENT by the Port Authority that Silverstein
Properties had been awarded the right to build a 1-million-square-foot
building on the last remaining site at the World Trade Center complex had left me elated and excited. The Port had already designed
the building, put foundation footings into the Con Ed structure, and
would collect monthly ground rent. The rest would be up to me. I
had to finance, construct, and then lease Seven World Trade Center.
And I was raring to go. I was confident I'd be able to move forward
quickly. This was a great location for a commercial building—close

to subways and the downtown financial district, and the Twin Towers were just a short stroll south. Tenants would flock to it.

But there was an important first step before I could start construction. While I had been awarded the right to put up Seven World Trade Center, that agreement was contingent upon my working out a ground lease with the Port for the site. We still had to agree on the terms of what the annual rent would be.

I had spent my entire adult life making deals and had learned a couple of lessons in the process. For one, a deal will only be consummated if both sides get something out of it. If the terms are disproportionate, if either the buyer or the seller is determined to make out like a bandit, it's never going to work. It's a certainty that things will eventually fall apart under the weight of unreasonable greed or arrogance.

This insight had led me to a second practical realization: you need to be flexible, willing to compromise. You can't just make demands and then expect the other side to acquiesce. Hardheaded is a recipe for disaster. If you come up against someone who is inflexible, out to have things his way and only his way, then you'd probably do best by calling the whole thing off.

And that was how I was starting to feel as the complicated ground lease negotiations continued to drag on with the Port for nearly a year. The Port Authority wanted things their way, and it was a very particular way to boot. Compromise was apparently not part of their vocabulary. Time didn't seem to bother them, either. Unlike those of us in the private sector, they didn't worry about deadlines, about negotiations dragging on as pre-construction costs continued to mount. They didn't seem to worry that there might be no building to rent for years and years; a seemingly endless flow of money from the New York and New Jersey airports, bridges, and ports kept pouring into their coffers.

In retrospect, perhaps I should have realized what I was getting into by doing business with the Port. And I certainly should have had a warning about how complicated a public–private sector business partnership can be. But at the time I was motivated by the opportunity to build a big commercial building on a very attractive site. The prospect was too exciting for me to be deterred.

There was also something else goading me on: I am a very tenacious guy. When I start something, I am determined to finish it. I had not won the lease only to walk off in a huff during protracted—and often infuriating—negotiations. Besides, I am an optimist. I believe that if I make up my mind to get something done, I will. In the end, I believe things will work out.

And so rather than giving up in exasperation, I instructed my people to continue negotiating with the Port. Then when things seemed to be getting out of hand, when the back and forth looked like it would go on forever, I stepped in. I set a deadline for finalizing the lease. If I didn't, the negotiations, I feared, would have gone on forever.

Finally, after more than a year, the deal was closed in 1980. At last I had my lease, and now I went to work.

STRAIGHT OFF, I HIT a home run—or so I thought at the time.

One of the guiding principles of the real estate development business is that in order to get the money you need to build from a lender, you must show that your project will be a profitable venture. And to do that, you need a tenant. So I went off looking for one. And pretty early on, I found one. Even better, he wanted more space than I'd ever imagined.

I had known Jim Boisi for years. When I'd first met him, he was working for the New York Central/Penn Central Railroad. The railroad owned the land along Park Avenue in Manhattan heading north from Forty-Fifth Street; their trains ran on the tracks below. This was incredibly valuable property that was sitting more or less vacant until Jim came up with an idea. He offered developers the chance to lease the air rights on the parcels of land above the railroad tracks. This allowed high-rise office buildings to be constructed along this stretch of Park Avenue—and at the same time gave Penn Central a huge source of annual income. And Jim was the guy responsible; his vision set it all in motion.

When J.P. Morgan saw what Jim had accomplished at Penn Central, they decided they wanted him on their team. He went to work for the bank and had a meteoric rise; he was very skilled. When I met with him in 1980, he was vice chairman.

I started telling Jim about the building I wanted to erect at the keystone site, and then showed him the preliminary plans for a forty-story structure.

After a careful look, he said, "The floors are too small."

"What do you mean 'too small'?" I challenged. Each floor was 25,000 square feet. This was a big building; there was 1 million square feet of rental space.

"I could use bigger floors," he said.

That got my attention, all right. "What size floors do you need?"

"How large is the site?" Jim asked.

"Fifty-three thousand square feet."

"Then give me a fifty-three-thousand-foot floor," he said.

"I've a million square feet here!" I countered.

"I could use two million feet," he said emphatically. "Give me a forty-story building with fifty-thousand-square-foot floors."

"That's doubling the size of the building," I pointed out.

"I can use it," he said confidently.

That certainly got me excited. Here was a prospective tenant telling me he wanted a building twice the size of what I had been intending to construct. Which meant I wouldn't have to worry too much about leasing: I had a single tenant who was interested in occupying the entire building. And the bigger space would mean I would be collecting significantly more in annual rental fees. "Okay," I told him. "Let me see what I can do."

So I went back to the Port and explained that I had a potential user for Seven. Only he wanted a building that was twice the size of the one we had envisioned. "What's the largest size of the floors I can build on that site?"

They looked at me like I was crazy. And maybe I was: doubling the size of Seven would be an enormous undertaking in terms of costs. It would also create new engineering and design problems; the keystone site was irregularly shaped, and there would be a limit on the size of the building that could fit on the plot.

But at the same time, the Port could do the same math that I had done. I was leasing the air rights from them for $6.75 a foot. Doubling the size of the structure would double the leasing rent they would receive. It would be a very good deal for both of us.

However, they quickly pointed out, there was one big obstacle to this potential windfall: there was an existing Con Ed substation that supplied electricity to most of Lower Manhattan that would need to be at the base of the office tower. The Port had already drilled holes to support pilings into the substation foundation, but they were engineered for a much smaller tower. Now they wondered if I would be able to sink the pilings and the enormous caissons necessary to support a new larger building into the bedrock beneath the operating electrical substation.

"I don't know," I answered truthfully. "But first things first. Is a two-million-square-foot building something that makes sense to you?"

They agreed that it certainly did make good economic sense.

"Okay," I replied with confidence. "Then I will figure out how to get it done." As I said, I have always been an optimist.

So I had my in-house design team get to work on some preliminary feasibility plans for a new, larger tower. Then I went to Emery Roth & Sons. They were the architects of record for the Twin Towers (Minoru Yamasaki was the design architect), and I knew the Port would have confidence in their work. And I did, too.

Emery Roth designed very efficient buildings, ones with, as we developers say, tight cores. That is, the elevators, stairs, and lifesaving systems were efficiently congregated in a narrow core space. And that kept all the surrounding space free for offices—which meant there was more room to lease. The loss factor, to use another term profit-driven developers are often bandying about, would be very low. And they were also skilled at keeping down costs, and that, too, was very much on my mind.

My instructions to them were simple: design a building with as big a footprint as possible for the site. Give me 53,000-square-foot floors, if you can.

They couldn't. The best they could do given the requirements for the airflow needed to ventilate the Con Ed substation was 47,000-square-foot-floors. But their design increased the number of floors from forty to forty-seven. And so I now had a plan for a building with 2 million square feet of space for the keystone site.

I hurried off with the plans to Jim Boisi. "Will this work for you?" I asked hopefully.

"This is terrific. Really terrific," he gushed. "We could use the whole building."

The whole building! This was unbelievable.

Except now I had to get Con Edison to allow me to plant test pilings into the bedrock beneath their substation. If I couldn't do these feasibility tests, then I'd never know whether I could drill down a hundred feet to submerge the caissons, each four feet in diameter, that would be required to support my tower. And I realized getting Con Edison to agree to this probe wouldn't be easy. They had a problem with me. And, I feared, it was one that might very well put an end to any discussions before they even started.

MY HISTORY WITH CON ED was this: I had been a joint venture developer in a partnership that had just completed the redevelopment of a building on West Forty-Second Street. It was a big building, 900,000 square feet. We'd bought it from a bank that had foreclosed, and when we took title the structure was in pretty bad shape. It needed a

Reputation Builder

Syndicator Larry Silverstein is betting that Seven World Trade Center will make him a Major Developer, too.

BY ELLEN RAND

Silverstein: "This has taken a bloody ton of staying power."

EVEN IN THE stratosphere where Manhattan's real-estate rich have their penthouses, there are layers of influence and layers of affluence. No matter how much money or status New York's real-estate developers achieve, there is always, so to speak, a floor higher to go. The need to have a new project never seems to fade. The Milsteins, in their early sixties, are building like crazy; so is Harry Helmsley, over 75.

Larry A. Silverstein, 53, hasn't built anything in Manhattan. He bought and renovated enough buildings in 30 years to control 7 million square feet of Manhattan office space, and he is worth at least $150 million as a result. But now, despite all that, Larry Silverstein is only looking up, reaching for a higher echelon.

He is telling a story about 521 Fifth Avenue, a 400,000-square-foot property where he has his Silverstein Properties headquarters. The ground-floor tenant went bankrupt, but before he left he began selling odd-lot merchandise, which he advertised heavily in the tabloids. Silverstein found the resulting crowds, not to mention the quality of merchandise, horrifying. "It certainly wasn't good for the image of the building," he says, literally shuddering in recollection. "It looked like the Lower East Side." Ultimately, Silverstein bought out the tenant's lease and replaced him with a

22 MANHATTAN, INC.

more dignified IBM Product Center.

Silverstein himself hails from the Lower East Side, many, many echelons back. This is where he first made a mark in real estate, pooling small investors and buying lofts to convert and lease. It's called syndication, of course, and Silverstein is one of the kings of it. After successes on the Lower East Side, he ventured into midtown, and the investors got bigger. In fact, Silverstein's estimated $150 million fortune is actually a controlling equity in some $700 million worth of properties.

Silverstein remains one of the few major players in Manhattan who has never built anything from scratch or by himself.

There is nothing wrong with that, of course, but ego is as important a factor in Manhattan real estate as anything else. Now, after 30 years in the business, having achieved social prominence by steadily climbing the hierarchy of leadership in the civic and philanthropic world, Larry Silverstein is aiming for Major Developer status as well.

His target: Seven World Trade Center, a 1.8 million-square-foot, $300 million office tower being built *from scratch, unsyndicated,* which, if successful, would be an echelon lifter if there ever was one. If it fails, some say it could cost Silverstein $50 million and a great deal of his reputation.

Silverstein made his first big leap only seven years ago, after he and ex-brother-in-law Bernard Mendik broke up their partnership, and he founded Silverstein Properties. Since then, the company has "grown like Topsy" in Silverstein's words. He's telling the truth. Among his Manhattan office acquisitions since 1978 are 120 Wall Street, 120 Broadway, 11 West 42nd Street, 530 Fifth Avenue, and 529 Fifth Avenue.

Now, Silverstein no longer syndicates his properties. When he needs cash he borrows from corporate pension funds, rather than taking in investors. As a result, his average equity in projects is increasing. Plus, the desire to do new developments for higher stakes has overwhelmed.

PHOTOGRAPH BY MARIA ROBLEDO

Larry Silverstein stands next to an architectural model of the original Seven World Trade Center.

complete renovation. As we went to work, we began to think about an energy plan that would make economic sense; the cost of fuel had spiked dramatically in recent months. We also wanted a heating and cooling system that would be environmentally responsible. Those considerations had led us to decide on a cogeneration plant for the building.

A cogeneration system works to combine both a structure's heating and electricity needs. Normally when you burn fuel for heat, you're only consuming between 30 and 35 percent of the fuel's energy; the rest goes up the stack, released and wasted. And polluting. With a cogeneration system, a significant portion of the energy that would've usually disappeared up the stack as waste is converted to steam to heat and cool the building.

Therefore, rather than hook into the city's customary electric grid, we used diesel electric generators to produce electricity for the building's tenants. Then we took the heat created by the diesel engines driving the generators and used it to convert water into steam. The steam was used for heating in the winter, as well as driving the air-conditioning compressors during the summer. It was a remarkably efficient system. Instead of 65 percent of the fuel's energy going up the stacks as polluting waste, we were able to put nearly all the fuel to effective use. It produced heating, cooling, and electricity.

The economic advantages were also impressive. It cost us $4 per foot to set up the cogeneration system, but the annual savings came to $1 per foot. That was a 25 percent return on our investment. Plus it was fully depreciable; the federal government, eager to encourage a nonpolluting, environmentally friendly system, allowed us to write off our investment in this electric diesel generator system against taxes. That was a tremendous savings, too. We got our money back in four years. A win-win for us: we helped the environment and we saved money.

Only Con Ed was furious. They claimed that since we were producing electricity we were acting as a public utility. And they sued us. We argued that we were simply providing a service for our tenants. We had no further ambitions. We weren't setting up a public utility company; we weren't doing business with anyone else. The judge listened and sided with us. Which only exacerbated Con Ed's fury.

How would this affect their willingness to agree to my conducting test borings into their substation on the keystone site? I didn't know, but I was apprehensive. When I went to my meeting with the Con Ed chairman, Chuck Luce, I made sure that I was accompanied by a representative from the Port Authority; I was hoping the Port's imprimatur might work to assuage any lingering bad feelings.

It didn't. As soon as I walked into Luce's office, he thundered, "You're the Silverstein who . . ." And then he went through the entire cogeneration lawsuit.

I pleaded guilty as charged, and I made sure not to reargue the case. My reticence seemed to help the situation cool down.

"You're not going to put one of those cogeneration things in this building?" he finally challenged.

"No."

"Will you put that in writing?"

"Sure," I agreed.

But Luce was not done. "You know, Larry," he continued, "you can't afford to fail. You do something down there that's going to cause us grief, power's going to go down in Lower Manhattan. And Con Ed can't afford to fail either. You're going to have to guarantee that there won't be some sort of outage."

"*Guarantee?*" I nearly shrieked. "Who can afford that? There isn't enough money in America to guarantee the power to Lower Manhattan."

We went back and forth for a bit. But in the end, as long as I signed a document confirming that Silverstein Properties would not have a cogeneration system in Seven World Trade Center, Con Ed, to my great surprise and delight, agreed to my request. To this day, I still don't know why Con Ed gave in. Curious, I have even recently asked them to search their files for an explanation, but they found nothing relevant. And there's no one still alive who could provide firsthand intelligence. But whatever the reasons, I'd gotten what I wanted. We could now proceed with our test.

I WAS AFRAID THE lights would go out on Wall Street. I worried, in fact, that all of Lower Manhattan would suddenly go dark, that there would be no electricity anywhere downtown. But I had no choice. I needed to discover whether piles that could support a forty-seven-story, 2-million-square-foot building could be driven beneath the

Con Edison substation that would serve as the base of the tower I now hoped to build.

The problem was that the existing electrical substation at the keystone site was crowded with very delicate, very sensitive load-shedding gears, as well as 110,000-volt transformers. The space between these devices was very constricted, very narrow. Barely a foot, in fact. Slam into any of them, or even give them a jolt, and no one could predict for certain what would happen next.

Complicating this further, the two-story Con Ed substation had been designed in 1967 to carry the weight of a forty-story, 1-million-square-foot building. Only now, over thirteen years later, we would be attempting to thread this delicate needle by inserting construction piles that could support a skyscraper nearly twice the size of what had been originally planned—with the fate of the electricity that powered the financial center of the world hanging in the balance. Yet if I was going to be able to move forward with my plan for Seven World Trade Center on the keystone site, I needed to find out.

There were, of course, several preliminary meetings before the day of the test. A team of our engineers as well as Con Ed's filled the room. Everyone was focused on the big challenge: How do we get this done while also keeping the Con Ed substation fully operational? How could we make sure that power would not be lost, even for a moment?

The sessions were very tense. The stakes, after all, were enormous. And at one of the meetings I turned to our chief structural engineer, Irwin Cantor. While it's architects who design the buildings, it's men like Irwin who make sure the buildings can actually be built. They make sure that the structures will stand up against wind, rain, and everything else nature and human events can throw at them. Irwin's a very competent guy, and a very decent guy to boot, but I was nervous, and he could see it.

"How do we do this, Irwin?" I implored. "The space is so tight."

So Irwin, trying to bring some humor into a very anxious situation, jocularly suggested, "Listen, we'll just have to get some Chinese guys to slither in . . ."

And then Irwin stopped in midsentence because he was suddenly aware that standing next to me was Jimmy Chan, a Chinese engineer on the Con Ed staff.

". . . or some Irish guy, or an Italian," Irwin added quickly. But it was a transparent attempt at recovery. All he could do was look at the Chinese American engineer and plead, "Jimmy, I'm so sorry. I apologize."

But was it too late? I was thinking this bit of wrongheaded, derogatory banter might very well cause the entire Con Ed team to storm off in an indignant huff. And I wouldn't blame them if they did.

But Jimmy was forgiving. "Irwin," he said, "it's okay. It's just a bad joke. We all make them. I know you didn't mean anything."

Then the day finally came for the test, and we began the insertion of the first test pile very slowly, all the time fearing something terrible might suddenly happen. All the Con Ed engineers were watching, waiting, looking to see if the pile could be inserted without making contact with the load-bearing gears or the transformers. If the Con Ed engineers saw something that didn't seem right to them, on their signal, we would have no choice but to stop the test.

But it worked. The pile went in smoothly on the first try. It was all handled effectively and efficiently. The lights were still on. It felt like a miracle.

And now that I knew a 2-million-square-foot building could be erected on the site, I had to try my hand at creating another miracle—raising the money that would be needed to get it done.

THREE

A N INVIOLABLE LAW OF real estate development is that you can't erect a building without having the money to pay for it. Yet as I prepared to go out to lenders who could provide the $300 million I'd need to build Seven World Trade Center, I quickly realized I had a big problem: it was a terrible time to raise funds. Actually, it was a lot worse than terrible; it was impossible. In the early 1980s the Federal Reserve, hoping to rein in inflation, had made borrowing very difficult; interest rates had risen to a dizzying 18 percent. At that cost, a viable deal—one that made economic sense for a developer—could not be made. And the banks realized that. None of them would even talk to us.

I decided I would just have to wait until the economy improved, but I had signed an agreement with the Port specifying that construction on Seven would be underway by 1982. There now was no way I could do that. Instead, there was the distinct possibility, I feared, that they would cancel my lease, claiming I'd breached the contract we had signed. Fortunately, though, I was able to convince the Port officials that the state of the economy was beyond my control and that it would be in all our interests to delay construction for a while. What good is putting up a building, I'd patiently explained, if you can't lease the space at rates tenants would find attractive?

So we both waited; there was really no other practical choice. As I bided my time, I kept telling myself that things would eventually improve. A day would come, I predicted, when borrowing would be a lot easier. As I said, I have always been an optimist.

Two long years later, in 1984, I was proven right. Things began to look more promising and I approached the Chase Manhattan Bank about borrowing the $300 million I needed to build Seven.

Their first question was, Do you have a tenant?

Sure, I replied confidently. J.P. Morgan.

Terrific, said the Chase bankers. Let's see the plans.

So I showed them the plans for the huge building with its 47,000-square-foot floors. They studied the drawings and, after not too much deliberation, decided, This looks like a deal we could be interested in.

I hurried to Jim Boisi at Morgan to lock things up. We'd reached the point when it was time to start negotiating the lease and work out the details of our financing with Morgan. After the years of waiting, I would soon finally move forward with construction. I was excited.

Jim wasn't. "We have a problem," he announced.

"What?" I inquired, my voice rising with a sudden trepidation.

He explained that his bank was troubled by the recent increases in city and state taxes. "We can't work in this kind of environment." Then he added, his tone resigned yet determined, "We're going to have to pull out of New York."

"Pull out of New York?" I echoed in bewilderment. "You've got your headquarters here."

"Well," he explained with a lofty impatience, "we've spent the past two years negotiating with the Koch administration to work out a tax deal that made economic sense. But in the end we couldn't get what we wanted. So we're taking the operation we'd planned to put in your building and taking it out of the city. We're moving it down to Delaware."

With that revelation, I lost all pretense of self-control. "Delaware?" I wailed. "We've got a great building for you here."

"Not going to work," he said with a banker's well-practiced stoniness.

I left his office feeling as if I were going to die.

WHAT DO YOU DO when all your plans fall apart? When years of work, when years of financial investment and considerable effort don't come to fruition? My lifetime in the real estate business has taught me that there's only one suitable response: you dust yourself off and try again. It helps, of course, if you can believe as I always do that things will work out in the end. Self-confidence, I contend, is a necessity for success in business.

So I returned to Chase. Look, I began sheepishly, I don't have the tenant I thought I had. But I'm still determined to build Seven World Trade Center. And I'm still set on its being a forty-seven-story building with 2 million square feet of rentable space. (Now that I had the design and had done the tests that confirmed that a huge tower could be constructed on the keystone site, I wasn't going to back off; the economics were too enticing.)

But, I continued, I also still need to borrow $300 million to get this done. Can you give me the construction financing? I boldly asked.

The Chase bankers responded by saying all sorts of encouraging things: we have faith in you; we believe that the huge open floors you've designed are terrific; you should be able to find Wall Street tenants for a building like that.

Then they added the "however."

However, said the bankers, since you don't have a tenant signed for at least part of the building, there's only one way we can move forward with the kind of commitment you want: you need to get us a takeout loan.

A takeout loan is a loan from a third-party financial institution agreeing that they will step in to pay back the construction lender if, when the building is completed, the developer is unable to make good on his commitment. These sorts of loans are structured like mortgages and are collateralized by assets. And they cost a pretty penny—all money that comes out of the developer's pocket. Think of them as insurance for the lender: even if the builder goes belly-up, the takeout loan will still cover the original loan agreement. Only the developer is paying the premiums.

Chase's financing, therefore, would be covered; they would have a guaranteed source of repayment even if I never found a tenant. Which was a good deal for them. But it would mean that I would be in debt for two loans instead of one. I would owe Chase its $300 million, plus I would owe whatever the takeout loan would cost. In addition, as I was well aware, I would be borrowing this fortune to build a tower where there was no firm assurance of my ever being able to rent the space. In the end, I could wind up with a shiny new building with 2 million square feet of empty space. And without tenants to lease the offices, how would I ever be able to pay off either of the loans? It was a question that kept me up at night.

But, I also told myself, if I could get Seven built, and if I succeeded in renting it out, I would make a good deal of money. Encouraged by this prospect, I began to search for an institution that would give me a takeout loan.

Back in the mid-1980s, when the seeds of the financial debacle that would soon be Wall Street's comeuppance were being planted, there were many financial institutions doing a thriving business in takeout deals. They blithely entered into them because in those high-flying times it seemed impossible that a developer wouldn't be able to find rent-paying tenants to provide the income needed to service the underlying initial financing. These institutions felt assured they'd earn money simply by providing window dressing, a form of insurance that would never be redeemed, so that the construction lender could hand over large sums without worrying too much. They were convinced it wasn't a risky business. They believed the day would never come when they would have to bail out the developer and pay off the primary lender. A disaster like that was an impossibility in these booming times. Or so they thought.

We soon found a California bank that was doing a brisk business in these sorts of loans. We had to pay them $7 million, and in return they guaranteed that in the worst of all circumstances, they would repay Chase the $300 million. Although I wondered whether they would have the resources to honor such a colossal commitment if several of their construction projects all incurred problems at the same time, Chase was satisfied. Once I had the takeout loan guarantee, Chase gave Silverstein Properties the $300 million I needed to build Seven World Trade Center.

And so we started construction on a 2-million-square-foot building. We didn't have a tenant. We didn't even have the prospect of a tenant. But we did owe $300 million to one bank. And had paid $7 million to another. People in the business were asking me, "Are you nuts?" I just smiled and told them I was sure it would all work out.

What can I say? I was young and foolish. And, as always, very confident.

BUT BY THE TIME the steel superstructure had risen to the third floor, I decided I had better start to look for a tenant in earnest. My chief candidates were the major downtown financial institutions, with the exception, of course, of J.P. Morgan; they had already moved to Delaware.

As part of this campaign, I made a call to Salomon Brothers, a Wall Street powerhouse. I gave them my usual pitch, telling them that I would like to show them the plans of the building we were putting up on the keystone site; I thought it might be a space that would appeal to them. And they said, Sure. Come on over. We'll

The original Seven World Trade Center, just north of the Twin Towers

be glad to meet. I couldn't help noticing that their tone was pretty unenthusiastic, but you never know.

So I brought my team to their offices and we met with some of their key people. I had started out in this business thirty or so years earlier as a broker leasing space, and now I was once again the salesman—only now what I was offering was not a loft for cutting fabrics but rather a 2-million-square-foot tower where the masters of the universe could wheel and deal billions of dollars. Still, the fundamental principles were pretty much the same: you need to show the prospective tenant why the space would work for them.

After I finished making the pitch, they announced, We're intrigued. How about you come back in two weeks and make a full presentation to the board?

The Salomon boardroom was packed as I, surrounded by my team, went into my presentation. We had come prepared, bringing reams of financial charts and building plans, but when I spoke it was from the heart. I told them what I had to offer was unique, and I truly believed it was. The building we had conceived with Irwin Cantor would have column-free floors. The bulk of the building's structural support, massive columns and huge fifty-foot steel beams, would be confined to the core. This meant that each of the 47,000-square-foot floors would be virtually column-free. When you looked down the length of a floor, there would be nothing but a vast, unimpeded space. And walls of floor-to-ceiling windows with heart-stopping views over Lower Manhattan.

The Salomon people imagined the picture I had painted with my pitch, considered it for a bit, and then decided, "The open space would be terrific for our trading floors. Just terrific."

It wasn't long before they announced, "We're in. We want a million feet." The next thing I know, we're spending weeks negotiating the lease, getting down to the brass tacks that are necessary before an agreement can be signed.

Everything was going along swimmingly. I was feeling pretty happy. And so was Chase, at last confident that their $300 million loan would turn out to be a pretty good investment for the bank. My long-held dream of erecting a profitable office tower on the keystone site, I felt, would soon become a reality.

WHILE THESE LEASE NEGOTIATIONS were playing out, another potential deal was vying for the attention of several major developers. It was for a prime uptown site. And it, too, seemed very promising. Yet I never suspected the impact this piece of property would have on concluding the deal with Salomon.

On Fifty-Ninth Street, just across from the southwest corner of Central Park, the site of the old Coliseum had recently become available. The building had been built for trade shows and exhibitions, but with the passing decades it had become antiquated. A new, much larger, and more modern convention center had been erected on the Lower West Side. The decision, therefore, was made to demolish the Coliseum and get the real estate community to bid for the right to acquire the site and develop it.

When the dust settled, Mort Zuckerman and his Boston Properties won. Mort headed down to Wall Street to confer with John Gutfreund, the man who ran Salomon Brothers. Forget about that lease you were negotiating with Silverstein, Mort urged. I've got a better building for you. Much better.

From a design standard, I had to concede that maybe he did: Seven World Trade Center was not conceived to be a showplace. There was no glitter, just function. It was a very efficient building whose purpose was to offer large, relatively inexpensive office space. The architects and engineers had succeeded in giving me a building that accomplished those ambitions, but it resembled, as one critic sneered, "an oversized shoebox." The assessment stung, but at the same time I knew I couldn't argue.

Zuckerman's building, the skillful design of Moshe Safdie, was something completely different. It looked really sensational, lots of spires, lots of doodads. And in addition to its architectural glamour, there would be breathtaking views of Central Park from the windows.

Gutfreund took one look at the plans and fell in love with what he saw. Then he called me.

"Larry," he began directly, "we're not going to make a deal at Seven World Trade Center."

My stomach began sinking. Then I was certain: I was dying. When I could finally talk, my argument was desperate. "You know you're going to be paying twice as much rent up there," I tried.

"Larry," he countered, "we're making so much money. It's not even a consideration."

As that sank in, he continued. "We need to be up there. Uptown. Salomon Brothers overlooking Central Park. It will be fantastic."

I felt like a jilted lover, the drab suitor thrown over for someone much more glamorous. But I knew from his voice there was nothing I could say or do to dissuade him. I had lost.

TO SAY I WAS low would be an understatement. I was building a forty-seven-story building that didn't have a single tenant. I owed the bank $300 million, without any real prospect for its repayment.

That evening, I couldn't eat. I couldn't sleep. Me, the man who was always so optimistic—all I could do was moan to Klara about my bleak predicament.

She listened without interruption. She let me get it all off my chest. And when I was done, she finally spoke.

"Listen, sweetheart," she said with a cool, unemotional detachment. "Something better will come along. I can't tell you what. But believe me: something better will appear. You just have to have faith."

And just like that, hearing Klara, so confident, say those words made them seem real in my mind. It wasn't necessarily logical, I concede. But it worked. Suddenly things seemed possible. I was once again the optimist.

"Okay," I agreed. "You're right. I'm going to have faith. Something better will come along."

KEEPING ONE'S FAITH, THOUGH, requires a lot of wishful thinking. Try being sanguine as you watch the steel go up on your $300 million building, only you don't have a tenant (or even the prospect of one). In those circumstances, faith can be a very demanding exercise. It was a good thing I had Klara to buoy my spirits whenever they grew shaky.

And as things turned out, Klara had been prescient. Not much more than an uneasy couple of weeks went by before, just as she had promised, my rekindled faith was rewarded. Out of the blue, Drexel Burnham Lambert, another Wall Street powerhouse, called. We would like to take a look at your building, they declared.

By this time, the building had been growing and I was able to give them an onsite tour of the steel-encased structure. Then I made a presentation to their board, emphasizing Gutfreund's insight that the huge column-free spaces would be perfect for trading floors. They listened, and said they would be in touch.

The call came a couple of weeks later. "Know what?" said the Drexel representative. "We like this building. We're going to take it."

"How much do you want?" I asked cautiously.

"The whole building."

"Two million square feet?" I reminded him. They certainly couldn't have been intent on leasing the entire building.

"Yes," the Drexel man said firmly. "We're going to take the entire building."

At that moment, hearing such remarkable news, I found myself thinking, *Klara had indeed been right all along! What a smart woman I had the good fortune to marry.*

TOPPING-OUT CEREMONIES HAVE LONG been a rite of the construction business. They traditionally occur when the final steel beam is placed atop the structure and are modest, self-congratulatory affairs. But when the building is a downtown Manhattan tower about to be occupied by a Wall Street firm employing thousands, the topping-out also becomes a media event. Drexel, in fact, wanted it that way and I, in truth, didn't mind. Attending the July 8, 1986, topping-out for Seven World Trade Center were Governor Mario Cuomo, Mayor Ed

New York Newsday
NEW YORK EDITION
WEDNESDAY, JULY 9, 1986

Finishing Touch on the Trade Center

Newsday / Richard Lee

Mayor Koch, developer Larry Silverstein, Drexel Burnham Lambert chairman Robert Linton and Gov. Cuomo raise beam yesterday.

By Ron Davis

The last steel girder was put into place yesterday atop the $300-million, 47-story office building that will complete the World Trade Center complex 16 years after it opened.

Seven World Trade Center, scheduled for completion this fall, will be the headquarters of Drexel Burnham Lambert, an investment banking and securities firm that plans to add about 5,000 employees to the complex's 50,000 workers. The red granite building located at the northern end of the trade center site in lower Manhattan will also add almost 2 million square feet of office space to the trade center.

The privately financed structure is being built by Silverstein Properties Inc., a real estate development and investment firm, on property owned by the Port Authority of New York and New Jersey. Owned by Silverstein Properties, the building will be leased to Drexel Burnham Lambert.

The new building takes up the entire block bounded by Vesey, Washington and Barclay Streets and West Broadway. Gov. Mario Cuomo and Mayor Edward I. Koch were on hand for a noon "topping off" ceremony in which the girder was hoisted by a crane and installed, and both said the new structure is a sign of a healthy economy in the city and state.

"New York City prides itself on being the financial capital of the world," Koch said. "Every new building that makes available more office space is welcome in this city."

"We're not just being swept along by some kind of economic recovery wave," Cuomo said. "The strength of the state and city is stronger than the national economic recovery movement."

Robert E. Linton, chairman of the board of Drexel Burnham Lambert, said half of the firm's 8,600 employees will occupy the building initially. He added that the firm plans to grow into the building and eventually have 10,000 employees there.

The World Trade Center is fully rented, with more than 1,200 firms and organizations representing a broad range of international commerce. It is made up of two 110-story office towers — which at 1,350 feet each are the tallest buildings in New York

City and the second tallest in the world — two nine-story office buildings, the eight-story U.S. Customhouse and a 22-story hotel, the Vista International, all constructed around a central five-acre plaza. The 16-acre complex is bounded by Church, West, Liberty, Barclay and Vesey Streets.

The first tenant moved into One World Trade Center, the north tower, in December, 1970, 10 years after development of a world trade center had been recommended by the Downtown-Lower Manhattan Association. The first tenants moved into the south tower and Northeast Plaza Building in 1972, the U.S. Customhouse in 1974, Four World Trade Center in 1977 and the Vista International in 1981. More than a million cubic yards of earth and rock were excavated to make room for the center. That material was placed in the Hudson River to create the 23 acres now being developed as part of the Battery Park City project.

Newsday / Richard Cornell

Mayor Ed Koch, Larry Silverstein, Drexel Burnham Lambert chairman
Robert Linton, and Governor Mario Cuomo at the topping-out
ceremony for the original Seven World Trade Center

Koch, and a big crowd of notables from business and politics. And I was the center of attention, or so it seemed to me. Look what you've done, people were telling me. You've built a 2-million-square-foot building and, even better, you've leased the entire space. Fantastic, Larry, everyone seemed to be saying.

Except I knew it wasn't fantastic—yet. The lease had not been signed. The negotiations with Drexel had been going on for over a year. There had been a lot of back-and-forth, a lot of give-and-take, but only now were the agreements ready for signature. We had scheduled the signing to take place three days after the topping-out. But I was pretty confident the signing would just be a formality. We had agreed on all the deal points. The Drexel people had helped orchestrate the topping-out shindig, had helped make sure that all sorts of movers and shakers would be in attendance. They were celebrating, too. What could go wrong in three days?

I soon found out.

My first clue that trouble was brewing came from the phones. They went silent. During the preceding six months, Drexel had been calling my lawyers, as well as my in-house team, several times a day. And many nights, too, had been punctuated with their incessant inquiries. But in the days after the ceremony, nobody heard from them. Even more disconcerting, when we called and left messages, no one got back to us.

The silence, I knew, was ominous. I decided I had better find out what was going on. My first call was to Herb Wachtell, whom I had known since 1944, when we both entered the High School of Music and Art as freshmen. Nearly a half century later, Herb was the senior partner in the esteemed Wachtell, Lipton law firm. And he was still my close friend.

Herb, I said, can you ask around and see if anything is going on at Drexel? I figured Wachtell, Lipton was involved in many deals, had their lawyers working on all kinds of cases throughout New York; they might be able to discover if something was up. Herb promised to make some inquiries and get back to me.

Then I called Larry Tisch. If anyone would be able to learn if something was amiss at Drexel, it'd be Larry. The co-head with his brother of the Loews Corporation, he was a businessman who had lots of wide-ranging investments, lots of contacts, and he was also a good friend; I served with him on the NYU board when he was chairman, and over the years of working together a genuine friendship had blossomed. He, too, promised to get back to me.

I must have made a few other calls, because it wasn't long before my phone started ringing off the hook. It wasn't Drexel calling, and it wasn't good news.

People began to tell me about Ivan Boesky. Boesky was a multimillionaire stock trader who had become enmeshed in an insider trading scandal. With the hope of cutting himself a deal with the authorities, he had starting telling the prosecutors about Mike Milken, a key Drexel executive. And now Milken, who had helped create Drexel's moneymaking machinery with high-yield and highly leveraged junk-bond financing schemes, was being targeted by the SEC for securities violations.

Forget about Drexel leasing a 2-million-square-foot building, people were calling to tell me. They won't be leasing anything. They won't even be a company for much longer. Drexel will never survive the Milken scandal. They were right. Drexel ripped up the lease before it was signed, and I was left with an empty forty-seven-story building.

The story of Drexel's demise and my predicament made it into the papers. It was a tale about two colossal failures, and I was reeling in the aftermath. This is not the kind of publicity I wanted.

After it appeared, I got a call from Lew Rudin. Lew and his brother Jack were two of the most successful realtors in the city. He was a terrific guy, a civic leader, a genuine cheerleader for New York. And another good friend.

"I know you think the world's coming to an end," Lew began, "but something good will happen."

"Oh my god, Lew," I groaned. "I have a two-million-square-foot building and it's one hundred percent vacant. I have a three-hundred-million-dollar mortgage that I have to pay off. I have a California bank backing me up, only there's no way they're going to be able to get that kind of money. Something good will happen? I don't see how."

"Trust me," Lew repeated. "Something good will happen."

Yet Lew, like Klara, was right. Something good did happen. And when it did, the person responsible was Jacqueline Kennedy Onassis.

MRS. ONASSIS STOOD IN Central Park with her open umbrella raised in the air, and about eight hundred other New Yorkers joined her, lifting their umbrellas, too. They were simulating the dark shadows that would be cast over the park by the two enormous towers Mort Zuckerman was planning to build on the Coliseum site. "One would hope that the city would act as a protector of sun and light and clean air and space and parkland," Mrs. Onassis said, and after she said it, many New Yorkers, including the *New York Times* editorial writers, agreed. It soon became clear that Mort would most likely not be able to build his proposed towers. And even if he somehow managed to succeed, John Gutfreund didn't want to be on the wrong side of an issue that was so close to the hearts of many influential and esteemed individuals. This was not the sort of environmentally insensitive image he wanted for Salomon Brothers, or, for that matter, for himself. There was now no way he could take the space in the glitzy Boston Properties tower with its sublime Central Park views.

When Gutfreund called, I could hear the hesitation in his voice. "Do you think you want to talk to me again?"

"I need to think about that," I replied with as much conviction as I could muster. But so much for my acting skills. I couldn't play the angry, spurned suitor for more than a few moments. "Yes," I agreed after a short pregnant beat, while all the time silently rejoicing. "I'll talk."

We talked throughout the next week, and by its end a deal had been negotiated. Salomon Brothers signed a lease for 1 million square feet, half of the entire building. And after this was announced, other tenants, eager to move to a gleaming downtown tower that had the imprimatur of a big Wall Street firm, signed leases, too.

Seven World Trade Center opened on May 9, 1987. Thousands of workers were soon arriving each day to offices in this red granite tower. And in spite of all the obstacles and setbacks, I had accomplished what I had set out to do. I had fulfilled one of my ambitions.

BUT I ALSO HAD another dream. I'd been silently harboring it since the day of the topping-out ceremony seven months earlier. I had been standing on the open roof of Seven, surrounded by the governor, the mayor, and a crowd of distinguished guests, and yet I found myself covetously looking across the way toward the Twin Towers. Sure, Seven was a big building, but I couldn't help noticing that it looked tiny, almost minuscule, when compared to the looming World Trade Center buildings across the way. They seemed to fill the sky. Each of the two towers was 110 stories high. Each was double the size of my building. Together they had 10 million square feet of space, and that was, incredibly, five times the size of Seven. They were enormous.

I found myself wistfully thinking, *Wouldn't it be something to own these towers someday?*

Looking back at that naïve moment, I can truly say that I had no idea I'd soon discover the wisdom in an old warning: Be careful of what you wish for—you just may get it.

FOUR

A REQUEST FOR PROPOSAL—AN RFP, as it is known—was a pretty common document in the real estate business. It was used by government entities to solicit bids on a project from qualified contractors or developers. An RFP, therefore, announced a competition. It was a contest run by the seller to determine who could meet its terms at the best price.

In 2000 the Port Authority issued an RFP that launched the contest for the biggest prize in the history of New York real estate—a ninety-nine-year leasehold for the two World Trade Center towers, as well as the considerably smaller adjacent buildings known as Four and Five World Trade Center.

What had happened was this: Not long after being elected in 1994, New York governor George Pataki made it clear that he wanted the Port to get out of the complicated business of running commercial real estate properties. With real estate valuations at a peak, the Port could reap, the governor suggested, a very attractive price by privatizing the sixteen-acre site, perhaps as much, the newspapers estimated (rather conservatively it would turn out) as $1.5 billion; money that could be spent to improve an aging mass transportation system.

The governor used his bully pulpit to persuade New York's representatives on the Port's board to support the sale and, after some additional cajoling, New Jersey's members also fell into line. In September 1998, the Port's board unanimously approved the idea of taking this iconic complex private.

But then (as I was to experience firsthand with increasing frustration over the subsequent two decades) a seemingly straightforward decision by the Port became embroiled in intrastate politics. For the next eighteen months the proposed sale was delayed as Pataki and New Jersey governor Christine Todd Whitman argued over the allocation of the hundreds of millions of dollars of spoils that would accrue from the privatization. Finally, in May 2000, the dispute was, at least in broad principles, settled. The following month the RFP went out to about thirty real estate development firms that the Port and its banking advisor, J.P. Morgan, considered capable of handling a deal of this size. They approached New York firms such as the Trump Organization, Vornado, and Tishman Speyer; national companies such as Mort Zuckerman's Boston Properties and the Rouse Company; and international firms such as Canada's Brookfield Properties.

And I got a call at my office from Governor Pataki.

"WOULD YOU HAVE ANY interest in responding to the World Trade Center RFP?" Pataki asked without much prelude.

I was somewhat surprised by the question. Silverstein Properties was certainly not one of the biggest companies in the business. And we certainly didn't have the financial resources of the well-capitalized and publicly traded real estate investment trusts (REITs). When *The New York Times* listed the major players potentially competing for the prize, Silverstein Properties hadn't even been worthy of a mention.

But I also knew we did possess something none of the bigger developers had—a good working relationship with the often willful and famously cantankerous Port Authority. And it wasn't just that the Port had been our (relatively silent) partner in the successful building and then profitable leasing of Tower Seven. In the horrific aftermath of the 1993 terrorist explosion at Tower One, when a 1,300-pound bomb had been detonated in the parking garage, killing six people and leaving both towers closed for months, I had quickly offered the Port interim office space in Tower Seven.

Great, they agreed. We'll get our lawyers working with your people on a lease.

Forget about the lease, I said. You need space now. Just move in. A handshake will work for the time being. We can work out the details later.

That was the way I had always done business. You treat people fairly, decently, in a manner you would expect them to treat you. When I give my word, whether it's signed on the bottom line of a contract or it's just a handshake, people know they have an agreement. My commitment is my bond.

When the Port moved their employees into temporary offices in Tower Seven on just our verbal agreement, they had learned how I did business. And years later they were still grateful.

As an implicit expression of that gratitude, as well as our shared success in bringing Tower Seven to fruition, the governor had called to inquire whether I would be interested in entering the competition to buy the Trade Center leasehold.

"Yes," was my emphatic and very immediate answer. "How could I not have any interest?" I added, although I didn't dare to confide my covetous thoughts about owning those two landmark buildings.

It wasn't, however, until I had concluded my call with the governor that I began to think about the prospect seriously. Was it actually a possibility? Could I compete against the better capitalized firms? Or was I fooling myself? Was my owning the World Trade Center complex just a pipe dream?

IN ANY POTENTIAL REAL estate deal, the first thing you want to do when you get serious is to look closely, with an unyielding objectivity, at what you might be purchasing. Sure, the prospect of owning the leasehold on the two tallest buildings in the world was heady; however, the bottom line of any deal is—the bottom line. You need to do the due diligence that will reveal whether the deal is worth pursuing, if it makes economic sense. You need to discover if your investment will in time turn a profit.

I began, therefore, the process of looking at the World Trade Center from this more pragmatic perspective.

First, I focused on the towers themselves. And what I saw, at least from a visual inspection of the buildings, was not very encouraging. When they'd opened in 1973, the two 110-story buildings, designed by Minoru Yamasaki, had not been very well received. Ada Louise Huxtable, the venerable *New York Times* architectural critic, had dismissed the hulking, uninspiring monoliths as "General Motors Gothic."

Over the ensuing decades, the buildings had not aged well. The lobby, a vast, cavernous space, struck me as disjointed, half a million square feet that was an unattractive labyrinth of shops as well as a confusing conduit to the belowground trains to New Jersey and twelve city subway lines. Entering the buildings, I felt, was like entering a maze. And, no less disconcerting, if you were going to an office on one of the upper floors, you needed to transfer at "sky lobbies" on the forty-fourth and seventy-eighth floors; that is, your commute would continue inside the building. That was the sort of tedious experience that could, I knew, discourage prospective tenants.

When I had bought my first small office building with my father nearly a half century earlier, one of the initial things we had done was to spruce up the place, bringing in buckets of white paint and sanding the loft floors. If I succeeded in gaining the leasehold on the

World Trade Center, I realized I would need to do basically the same sort of redecoration—only this time for a lobby the size of a football field that rose up thirty feet high.

That would be costly—tens of millions of dollars, I roughly estimated. And complicated; even spending generously, there would be no guarantee that I would succeed in solving all the aesthetic and logistical problems. I needed, therefore, someone whose expertise I could trust.

I called David Childs, a senior partner in the esteemed Skidmore, Owings & Merrill architectural firm, and another of my good friends. "David," I said, "if I get this, I want you to help me reinvigorate the Trade Center. Would you be on board with that?" David swiftly agreed, and I breathed a sigh of relief. I knew that if the time came to move forward, I would have someone whose taste and judgment I respected ready to jump in to redesign the perplexing and off-putting entranceway to the building. That, after all, was one of the first rules of commercial real estate: you need to give people a space where they'll enjoy the experience of coming to work.

Yet what about all the stores that filled the expansive lobbies? There was 430,000 square feet of retail space, and that added up to seventy-four separate stores and food shops. What did I really know about this complicated and competitive retail world, about attracting premier tenants, negotiating the complex leases? My experience was in commercial real estate, in developing and operating offices.

It quickly occurred to me that it would make good sense—and save me a lot of headaches—if I could partner with someone who had this unique sort of knowledge and experience. As it happened, I had a good friend who had precisely this sort of expertise. Frank Lowy had managed to survive in a Hungarian ghetto during World War II, emigrated after the war to Palestine, fought valiantly in Israel's War of Independence, and later went to Australia to be reunited with his mother and sister. In 1959 he opened his first shopping center in Sydney. Forty years later his family-run Westfield Corporation was one of the largest mall operators in the world. And after several conversations, as well as undoubtedly a shrewd bit of his own due diligence, Frank agreed that he would join me in my potential bid; they would create a Westfield shopping center in the 430,000-square-foot Trade Center retail complex. The actual extent of their contribution to our joint final bid still had to be determined, and in the months ahead, as in any business deal, there would be some tough give-and-take, but for now at this preliminary stage I was confident that I had effectively dealt with this concern.

Nevertheless, there still remained one crucial question: Even if I could spruce up the tired old towers, even if the stores were no longer my problem, could I charge rents that tenants would find reasonable and yet would still allow me to make a profit? The towers would cost a small fortune—several billions, I estimated—to buy, and I wasn't going to plunge into what would undoubtedly be a long, demanding, and expensive competition for the leasehold out of the sheer vanity of owning what were then the two tallest buildings in the world. I wasn't in business to lose money.

And so I began doing the math.

I started with the office space—and just a rudimentary analysis was an education. Sure, there were 10 million square feet, and that translated into a daunting 450 separate leases; the towers were 98 percent rented. However, I discovered that 74 percent of the space—7.4 million square feet—was occupied by just forty tenants. And these tenants, I further learned, were paying an average rent of $32 a square foot on leases that were due to expire within five to six years. Could I charge a higher rent and still keep these tenants, as well as attract new ones?

I didn't have to do any research to answer that question. I had recently acquired another downtown office building, 140 Broadway. It was about half the height of the Twin Towers, a mere fifty-one stories, and not long after I took control, a large chunk of office space had been vacated by a foreign bank. Starting at the second floor, twenty floors were suddenly available, and none of the views were particularly noteworthy; the lower floors, in fact, looked south directly into the brick and granite façade of another building I owned, 120 Broadway. Yet I had succeeded in relatively short order in signing a twenty-year lease with an investment concern for the entire twenty floors—at $50 a square foot.

Now, I had seen the views offered by the World Trade Center offices, and even from the lowest floors they were a marvel. You could look out into the unobstructed distance, the city and the Hudson River spread out magnificently in front of your eyes. I had no doubt that I could rent these offices with their open, breathtaking vistas for at least as much as I was receiving for nearby Wall Street space that stared out at a grim building façade.

You don't need to be a math whiz, or a particularly astute real estate developer, to grasp that between the $32 a square foot rental the Port was getting and the $50 I firmly believed a new owner could soon charge when most of the leases in the Twin Towers expired, there was an $18 differential. And when you multiply that increase in rental income times an eventual 10 million square feet of space, well, the upside to this was enormous. *Huge.* The increase in rental income would potentially be more than $180 million each year.

But that was not all. I had spent most of my adult life putting together a company that knew how to run commercial buildings. We understood how to get things done efficiently, and at the same time we kept careful track of every dime that was spent. The Port did a good job in running the towers; however, the bottom line was never their primary concern. Government entities, after all, don't have to worry about making a profit. The rental staff for just the Twin Towers, for example, was larger than the number of employees I had in my entire company, and I had dozens of buildings. The Port's cost for operations at the Trade Center came out on average to a pricey $14.50 a square foot. I knew Silverstein Properties could get the same job done at a cost of about $9 a square foot. And so here was another big savings: $5 a square foot. And when you multiplied that by 10 million square feet—well, the potential for profit increased by about $50 million each year.

The Port also, I found out, for some arcane reason didn't measure space by the same procedures that most private landlords in New York—including Silverstein Properties—used. And when I applied the standard real estate measurement system to the towers' vast footage, the space I would be renting grew even bigger. The adjusted footage could bring in an additional 6 to 8 percent in overall annual rental income.

Add all these pieces together—the increase in rents, the lower operating costs, and the additional rental footage—and the upside on leasing the complex was enormous. If I could purchase the leasehold at the right price, there was the potential to make a fortune. The profit over the next six years, I estimated, could exceed several billion dollars.

But there remained one big sticking point before I could even begin to contemplate entering the chase. To qualify as a bidder, a company had to convince the Port and their banker that it had the financial resources to handle a transaction that would, it was growing increasingly clear, have a price tag of several billion dollars. Silverstein Properties didn't have that kind of money. I would need, therefore, to find a financial partner that did.

To my astonishment, it didn't take me long. I went to GMAC, a deep-pocketed lender (although, in time, it would become another victim of the financial crisis precipitated by the collapse of the subprime mortgage market), and ran through the broad outlines of the sale—what it would take to close the deal, and what the upside could be for an investor. I must have given them a very convincing pitch because they soon gave me their commitment to securitize the purchase debt. They agreed to put up at least two-thirds of the actual price, with the remainder coming from myself and other investors. And, more good news, I somehow got them to agree that they would exclusively back my bid; they would not consider aligning with any other potential buyers. This certainly gave me a major advantage as I moved forward. And, sure, their commitment to backing my purchase was very fluid—"semi-firm" is the term we use in the real estate business—and I anticipated many difficult conversations in the days ahead as they would undoubtedly demand significant guarantees for their funding. However, I now had the financial partner I felt I needed to convince the Port that I was a legitimate suitor—someone who, if selected, could hand over a significant down payment at the close of the transaction.

There remained one nagging worry, and it had the potential to send all my heady estimates of profit crashing into disarray—the PILOT agreement. This acronym stood for "payment in lieu of taxes," and it was a huge concession that New York City had given to the Trade Center when it had first been built.

In 1973, the Port Authority was having a difficult time finding tenants for the towers. They had 10 million square feet of office space, but their initial tenants were largely government entities; private companies had found they could choose from a glut of downtown office space, all offered at cheaper rents. So the beleaguered Port convinced the city to give them a break on the real estate taxes—and suddenly the Trade Center had a distinct advantage in the rental marketplace. A company could rent the identical amount of space in the towers as it would in a Wall Street area office building, but at a lower cost. At the time I was looking at the complex, the annual city real estate taxes were about $25 million. But if the PILOT incentive was removed and the buildings were taxed at the normal commercial rate, the taxes due would immediately increase to $100 million. And that would just be the initial increase; taxes would very likely keep climbing year after year. An additional $75 million or more each year during the course of the ninety-nine-year lease would add up to a significant amount of money. Enough, arguably, to turn a profitable deal into a losing venture.

But, I asked myself, was there any real possibility that the PILOT agreement would be removed? It didn't seem likely, I decided. And I moved forward with great enthusiasm.

"If I can get this damn thing," I exclaimed to Klara, "it would be a home run and a half."

THE INITIAL ROUNDS OF bidding began. One problem was that the Port at this stage was only providing very limited leasing, operating, and financial information. And they wouldn't budge. Everyone had to submit their qualifications along with preliminary bids based on a review of very sparse intelligence. Another concern was that everyone's bid was supposed to be confidential. Details, however, kept appearing in the newspapers. It seemed like every time I picked up *The Wall Street Journal,* there was another leak. Apparently, some members of the Port's Board of Commissioners, eager to curry favor with reporters, had decided to ignore the confidentiality agreement. I complained to the Port, but it didn't matter. The leaks kept being published.

So all I could do was plow on. From the list of the thirty or so companies the Port had originally approached, twelve firms were short-listed. And Silverstein Properties was one of them.

There was one more intermediary cut, and, again, Silverstein Properties was left standing.

Then, in October 2000, the Port announced the finalists for what they called "the best and final bid." Four companies were selected. There were three large, well-financed public companies: Vornado, run by Steve Roth; Mort Zuckerman's Boston Properties, which would quickly decide to team up with another finalist, Brookfield Properties. And one small, privately owned company: Silverstein Properties.

We were, relatively speaking, the minnow in a sea filled with whales. And I'm not just talking about the size of our organization, although that was very much the case when compared to the other groups, that had hundreds of employees. What I mean is this: Sure, I would be the first to volunteer with pride that I had come a long way in the real estate business over the past four decades. I had a very nice portfolio of properties. But it's not false modesty for me to say that the competition were much bigger players in the field. They had amassed large inventories of very lucrative properties. And as publicly traded companies whose assets and credit sheets were pretty much a matter of carefully audited and readily available record, they had a big head start over a tightly held family business like mine— especially in the sort of negotiations with the banks and commercial lenders that would be a necessity in a deal this large. After all, the ultimate price tag would be billions of dollars.

Nevertheless, I moved forward believing that we possessed a few advantages. As mentioned, my years of experience in building and running Tower Seven had taught me how to deal with the often dogmatic Port. And for another, as a private company I was not under the legal constraints that would hinder a publicly traded company. In this bidding war, I decided, bigger was not necessarily better. I didn't have to open my books or justify my actions to any stockholders.

Maybe, I began to feel, we had a good chance: We could win.

AND NOW THE CLOCK was ticking. The deadline for the submission of final bids for the Trade Center was January 31, 2001. A mere four months—with the ownership of two 110-story towers, an additional two nine-story buildings known as Four and Five World Trade Center, six labyrinthine levels of subgrade space, and the 360-foot broadcast mast on top of Tower One going to the winner. Who, I was starting to realize, would need to write a check in excess of $3 billion. And adding further stress to an already time-pressed challenge, the Port sure didn't make it easy for any of the three finalists to get the information we needed to determine our bid.

All the relevant documents had been packed into cartons and stacked in a single room in the Trade Center. There must have been about two hundred cardboard filing boxes, and each was crammed with pages and pages of very complex legal contracts and agreements. None of the finalists were allowed into the room at the same time; access was granted according to a specific schedule set by the Port, and even then the allotted hours were limited. Photocopying of the pages was prohibited; all you could do was take notes. The Port, I felt, was deliberately making things difficult, forcing the suitors to prove their ardor by overcoming all obstacles thrown in their paths.

With so much to review, and so little time to do it, at this point I brought in a team of lawyers to coordinate the due diligence. Len Boxer, a senior partner at the Stroock & Stroock & Lavan firm in New York, was another old and trusted friend (that's how I do business: I usually work with the friends I have made over the years). He quickly assembled a group that eventually included sixty-five people—lawyers and paralegals—who were culled from eleven different specialized practices within the firm. In addition to those drafted from the New York headquarters, Len reached out to the firm's offices in Miami and Los Angeles and brought personnel to New York for the duration of the bidding process. It would be a massive undertaking: people would be working long days and well into the night, a virtually nonstop, all-consuming whirl of activity until the deadline. The legal bills alone would be in the millions.

Yet even before the process began in earnest, I remember Len telling me, one friend to another, "Larry, this is a pipe dream. You'll never win against the competition."

"I'm not giving up without a fight," I shot back.

THE STROOCK TEAM REALIZED that there was just not enough time to review every contract. There were, for example, 450 separate commercial leases, and each was a thick and intricate document written by shrewd lawyers. The first step, therefore, was to cut down the piles of scrupulously crafted tracts that the Port had provided to a manageable size.

We decided to concentrate on leases that were for one floor or greater—and that left us with forty-one contracts to analyze in depth. Now, while this certainly reduced the field, it nevertheless remained, given the rapidly approaching deadline, a formidable undertaking. The leases had to be examined from both a business and an accounting perspective. We had to understand the significance of what the rent roll was today and we also had to project it well into the future. For example, Morgan Stanley was a tenant with an existing lease for 1.2 million square feet, and that alone was more than 10 percent of the entire commercial space. The lease would expire in 2006; what if Morgan Stanley didn't renew? What if we couldn't find a tenant? But what if we could raise the rent significantly? All these possibilities, and more, had to be considered. In addition, we had to compute what the management fees and leasing commissions would be over time; in the real estate development business, the equity in the building is very often not the biggest source of potential income.

Yet the commercial leases, I have to emphasize, were only one part of the puzzle. We had to review hundreds of service contracts, and the Port had its own very specific requirements for maintenance and security. We had to analyze very sophisticated tax concerns; after all, I was not attempting to buy the complex but rather was negotiating a ninety-nine-year lease that required my paying an annual "rent" to the Port for the land beneath the buildings, and this created a unique tax situation. There were also a myriad of title and survey issues.

One particularly problematic area involved the "slurry wall," a solid, three-foot-thick belowground concrete barrier that surrounded the Trade Center site. It had been designed to hold back the Hudson River and to prevent the basement levels from flooding. The engineers had made sure this "bathtub," as the construction workers had called it, would stand up to pretty much anything. It was held in place by more than a thousand steel-cable tiebacks that were anchored deep into the bedrock below the subbasement of the Trade Center. Nevertheless, the slurry wall's continued reliability— something that had to be guaranteed yet was in truth difficult to

The "slurry wall" prevents water from the Hudson River from
seeping into the Trade Center site.

predict—had an effect on both ongoing leasing agreements and
potential insurance claims. And further complicating the issue of
insurance liability, there remained unresolved litigation stemming
from the 1993 attack on the Trade Center. What would be the legal
and financial exposure of a new owner of the leasehold to the still
lingering and very sizable claims for damages?

Then there were fifteen separate broadcast and telecom tenants,
each with complex legal agreements that spelled out the terms for
their use of the 360-foot antenna on Tower One. And this massive
antenna, too, came with its own insurance, maintenance, and upkeep
contracts, and these documents had to be analyzed as well.

Finally, and not least by any means, as in any real estate deal
regardless of its size, there was the bottom-line concern of money.
How much would we have to put up to close the deal?

THE PORT, AS I have said, was determined not to make things easy.
They required that even before the deal could be consummated—

that is, even before the deficit financing agreements with the lenders (in my case, GMAC) that would provide the major portion of the purchase price could be fully negotiated—the winner of the competition had to put up a deposit of $100 million. And going "hard" with that amount of cash before the deal would be consummated, well, this was a staggering sum, a requirement designed to separate the wheat from the chaff.

Further, I have to concede that I didn't make things easy for my people to raise this amount of cash. When every potential investor was approached, I immediately announced my non-negotiable credos: This was my deal. I had brought it this far, and I would bring it over the finish line. I would have the final say in any dispute. I would negotiate the final terms with the Port. And I would run the complex if our bid was chosen.

Nevertheless, even if some investors were willing to accept my single-handedly running the show, there was another issue they raised that I had to agree made some sense. They were handing over their money to a seventy-year-old man. What would happen if I wasn't around?

But in the end, after months of haggling and with Len Boxer's wise counsel, we put a plan together that would enable Silverstein Properties to hand over a $100 million letter of credit to the Port—if we won the deal. I, along with several of my longtime investors, would put up some of the money, while Len had brought in a group headed by Lloyd Goldman, whose family had deep roots in the New York real estate business, who committed to put up the remainder. (I remember in the early days when Lloyd first approached me, he'd be calling seemingly every day and asking if I'd like to meet. I tried to play it nonchalant, putting him off until finally saying, "I guess it couldn't hurt." And in the end, it sure didn't. Lloyd and the Goldman family became knowledgeable, straight-shooting partners whom I would come to count on.) All the parties agreed to put up an additional $5 million as a down payment on the ongoing due-diligence expenses.

As for the return on their investment, an arrangement was formulated that made promising economic sense. In broad strokes, it provided over time a 12 percent return on the equity contributions; then a return of capital; and after that we'd be splitting the profits fifty-fifty. And in addition to management and leasing fees, I'd also receive a payment for putting the package together. It would be a good deal for everyone. *If we won.*

THEN, JUST AS THINGS were really heating up, when the deadline for the submission of the bids was only six days away, I decided that I could use a break. It would be restorative to step back, if only for an evening, from the seemingly constant meetings with lawyers, accountants, and investors. And so on January 25, 2001, I went to the annual dinner hosted by John Cushman. John was the chairman of the board of Cushman & Wakefield, a major player in the real estate world, and everyone in the business would be there. Including all the other finalists bidding for the World Trade Center. I wanted to be there, too.

At seven, when Rocky, my driver, dropped me off at Le Cirque, I told him he might as well go home. He had a wife and children waiting for him. It was only a short six blocks from the Midtown restaurant to my apartment on Park Avenue. After a big meal, the exercise would do me good. I'll walk home, I decided.

It was the worst decision of my life.

FIVE

I T WAS A HOLIDAY tradition, the one party I looked forward to attending each year. John Cushman's lavish dinner at Le Cirque had everything going for it—a festive setting, good food, even better wines, and, not least, a chance to socialize with the most important players in the real estate business. The room was always crowded, and it was always buzzing.

This year as I made my way through the clusters of people, all the talk, I discovered, was largely focused on the same topic: the contest for the World Trade Center leasehold. Who was bidding what price? Who had an advantage? Who would walk away with the greatest prize in the history of New York real estate? It seemed as if everyone had heard a rumor or had an opinion.

I listened to all the conversations, but didn't say much. From what I overheard, though, it was pretty clear that no one thought Silverstein Properties had much of a chance. The price tag would be too big and the management issues too complex for a relatively small private company. The winner, as the conventional wisdom had it, would be one of the capital-rich REITs. I didn't bother to object. I kept my cards (and my plans) pretty close to my vest. And when John Cushman made his annual toast and this year offered each of the three finalists in the bidding process good luck, I simply smiled politely. I didn't want to give the impression that I was overconfident. Yet, as I have previously mentioned, I couldn't help thinking to myself that unlike my competitors, I knew how to work with the Port. And that a private company could do things, make accommodations, that a publicly held company can't—or simply won't. Despite the fact that all the knowledgeable people in the room weren't giving me a chance, I still believed I could walk away the winner.

But that was my secret. In the meantime, I would wait quietly—and then have the last laugh. Or so I hoped.

The evening ended with John giving all of his guests a present. He owned a vineyard in Santa Barbara County—wines were his passionate hobby—and he presented each of us with a jeroboam of his latest Syrah. But there was more to his gift than just the big bottle of red wine; it came with two crystal glasses and a corkscrew, and it was all packed in a very elaborate case. It was really a lovely and generous present.

Only it weighed a ton. After I put on my long, heavy winter overcoat and cradled the bundle in my arms, I felt like I would be setting off to traipse through the icy Manhattan night carrying a deadweight. I immediately regretted having sent my driver home. For a moment, I thought about leaving the parcel behind; I could drive by in the morning and pick it up. But how would that look? John had gone to all the trouble of organizing this gift of his own wine, and I would be refusing it? It would be rude.

So I cradled, as best I could, the substantial cloth package in the crux of my arm. And then I trundled off into the wintry night for the six-block walk home.

I WAS DEEP IN thought. I was playing back in my mind all the conversations at the dinner, and I was considering all the challenges that still needed to be analyzed and resolved in the few days that remained before I submitted my final bid, when I came to the corner of Madison and Fifty-Seventh Street.

Yet despite my churning thoughts, after a lifetime in the city I instinctively stopped to look up at the traffic light. It was green; other people were crossing. I headed across the street, too.

Wham!

A car sped through a red light. I never saw it coming. Yet it hit me with enough force to hurl me eight feet. All of a sudden I was lying flat-out on the street. I had never felt such pain in my life.

I kept going in and out of consciousness, but when I managed to look up I saw that people were hovering over me. And I could hear the concerned voices. "Don't move! Don't move!" they were saying. "An ambulance is on the way."

I don't know how long I lay on the cold asphalt street, but it seemed forever. Finally, there were two ambulance attendants at my sides trying to lift me onto a gurney. When they moved me, the pain was unbearable. I must have blacked out. When I regained consciousness, I was inside the ambulance, strapped down on the gurney.

I could hear the siren screaming as the ambulance headed off. It didn't go far before it hit a pothole. And then another. Each jolt rocked every jangling nerve in my broken body. Oh god, did it hurt. "Go slow," I begged. "Please."

The pain, however, pulled me back into consciousness and got me thinking once again. "Take me to NYU," I ordered. I had been serving on the NYU Medical Center board for several years. In the future I would help establish a fund that would pay the entire tuition and expenses of every student enrolled in the medical school during the entire length of their studies, including postgraduate training in specialized areas. It would give Klara and me a great deal of pleasure to be able to assist in the education and training of a new generation of doctors. But back then I was just hoping that in my time of need and distress, the NYU doctors would be eager to help a board member.

Only the ambulance driver wouldn't listen. The Hospital for Special Surgery is closer, he said. It's on Seventieth Street. Nearby. That's where we're going.

I realized there'd be no point in arguing. "Okay, wherever," I acquiesced. "Just please go slow," I urged, my voice a desperate plea.

Once I arrived at the hospital, they started moving me to the X-ray machine. Only as soon as they attempted to lift me off the gurney, I let out a howl. The pain was just terrible.

"We need to get you into X-ray," the doctors insisted. "We need to see if there are internal injuries. If there's internal bleeding."

I could appreciate their point, but I just didn't think it could be done. I wouldn't be able to bear the pain. Nevertheless, I girded myself for the X-ray exam. Doubting I would survive the ordeal, I made a few last requests.

I gave them the number of Roger, my son. Please call him. Let him know I'm in the hospital. And ask him to call Dr. Farber.

Dr. Saul Farber was head of the NYU Medical Center. I had worked with him for many years, helping to formulate the hospital complex's real estate decisions. I knew him well and trusted him. I thought he would be able somehow to rescue me from the clutches of this strange and uncomforting hospital.

I also gave them the number for Lisa, my daughter. Please let her know where I am, what has happened. And I made sure that someone called Klara, who was in California visiting our daughter Sharon; as soon as she got off the phone, Klara hurried to catch the first flight back to New York.

Meanwhile, I was being lifted onto the X-ray machine.

A long, cruel lifetime later, they finished. Then they gave me the news: Miraculously, there were no internal injuries. My pelvis was broken in twelve different places, but these were clean breaks, and this, they said, was encouraging. I wouldn't need surgery.

However, I wasn't in any mood for celebrating. All I wanted was morphine. Anything that would numb the pain. I waited impatiently as the intravenous drip was inserted into my arm.

At last I could feel the morphine easing its way through my broken body. And I drifted off into a deep, deep sleep.

"WHAT DAY IS IT?"

I was slowly awakening from my drug-induced sleep. And there was Klara by my bedside; I was so happy to see her. I wasn't sure, but I instinctively suspected a good deal of time had passed since I had gone under.

The doctor announced that it was January 27.

Despite my grogginess, I did the math: I had been sleeping for two whole days.

It took a moment for that to sink in. Then I realized: it meant the Trade Center bid would need to be submitted in just three days.

"Kill the morphine!" I ordered.

"You're going to feel terrible," the doctor warned.

"I know, but it can't be helped. Kill the goddamn morphine."

The doctor looked at me like I was crazy, but he nodded in agreement.

"And get me a room where I can meet with my people."

There was a private conference room I could use, he suggested.

"Great," I agreed. Then I reached for the phone and began calling my team.

"We're back in business," I excitedly announced into the receiver. "Come to the Hospital for Special Surgery. *Now!*"

MORPHINE, I SOON DISCOVERED, was not the only way to kill pain. Ambition proved to be another effective way to refocus the mind. A rapidly approaching deadline helped, too.

For sure, I wasn't feeling great and I couldn't walk; I was still flat-out in my hospital bed. But the hospital conference room was crowded around the clock with people from my office, the lawyers from Stroock, and my accountants. We kept analyzing all the issues, plugging numbers into the computers, trying to tie up all the loose ends. And all the time we were trying to settle on a number—the price that we could live with and that at the same time would allow Silverstein Properties to walk away with the prize.

On January 31, we submitted our bid to the Port: we offered $3.2 billion for a ninety-nine-year leasehold on the World Trade Center complex.

It was now out of my hands.

So I asked the doctor if he now could, please, give me something for the pain.

TWO DAYS LATER, THE hospital released me. I had been there for nearly a week. I still couldn't walk without crutches and, in truth, could barely move about with them. But at least I was going home. For a while, I had not been sure if I would ever leave that hospital. I was elated.

Once I was in the car, I told my driver, "Listen, do me a favor. Let's not go straight to the apartment. Let's just drive around a bit. Maybe head up toward the park. It'd be nice to see some trees."

Of course he agreed.

And I quickly added, "Just don't hit any bumps."

So we were driving around, and it's as if I'm looking at New York for the first time. All the people. All the hustle and bustle. I felt like I'd been reborn.

Then the car phone rings. It's my office. We have a call from J.P. Morgan for you, they said. Do you want to take it?

I knew why J.P. Morgan was calling. They were the bankers representing the Port. A decision must have been reached. But I hadn't even been out of the hospital for thirty minutes. Did I really want to get into this now?

"Sure," I told my secretary. "Put them through."

"Mr. Silverstein," began the banker at the other end of the line without preamble, "we have some bad news for you. You were the second-highest bidder."

A long moment passed in an uneasy silence.

I finally managed to speak: "Who won?"

"Vornado."

I hadn't expected that. Even *The Wall Street Journal* had predicted that once Boston Properties, Mort Zuckerman's REIT, had partnered with Brookfield, the successor to the international property development firm of Olympia & York, they would be the front-runner.

"What was the winning bid?" I asked, still numb from the news.

"Three and a quarter billion."

"Fifty million more than mine," I conceded. And I thought, *With those kinds of numbers, it amounted to little more than a rounding error.*

"What was the lowest bid?" I asked.

"Fifty million below yours," the banker answered.

I did the math in my head. The three bids were $3.15, $3.2, and $3.25 billion. We were all so close. But only one could win, and it wasn't us.

"We're giving Vornado twenty-one days to close the contract," the banker continued. Then without another word, he hung up.

And at that low moment I felt as if another speeding car had just slammed into me.

I HAD LOST. ALL the months of hard work, all the careful planning, all the millions spent on due diligence—only to lose in the end. And by a "mere" $50 million at that.

"Let's go home," I told Rocky, my driver. I was no longer in any mood to drive complacently around the Upper East Side. I wouldn't be buoyed by seeing the Central Park trees. In truth, at the moment I was feeling pretty sorry for myself.

But once I got into my apartment and saw Klara, I was able to put things in perspective, and my mood quickly lifted. Sure, after months of strategizing I had come in second in a multibillion-dollar bidding war. And, sure, I was disappointed. But, I reminded myself, look at all I had. Look at how lucky I was. I was back with my wife. Back at my home. A week earlier I hadn't known if I would live or die. And even if I managed to survive, there was no guarantee that I would ever walk again. All things considered, I reminded myself, I was a very lucky man.

"Klara," I told my wife, "we should take a trip. Once I'm up and walking again, we should go somewhere." Now that I wouldn't have to spend my days and nights feverishly working to close the contract on the Trade Center, I would have time to enjoy things. Maybe I would even slow down a bit. I was seventy, after all. If the accident had taught me anything, it was that there was no point in putting things off. You never know how much time you have.

I was happily mulling this revitalized way of looking at life when the phone rang. Phil Moskowitz, my internist at NYU Medical Center, was calling to check on how I was settling in.

"If you get a fever, if anything starts to bother you, call me," he instructed.

"Why should I get a fever? Remember—I just got out of the hospital," I teased.

"Look," Phil reiterated, "you get a fever, just call me."

"Sure, Phil. Will do." But I'm thinking, *Doctors. They never want to leave well enough alone.*

Except an hour later, I had the chills. I'm shaking—literally. So Klara found a thermometer and I took my temperature. What I read convinced me to call Phil at once.

"I'm coming over," he decided after I filled him in and had answered a few of his clinical questions.

"What do you mean, you're coming over? You're too big a deal to do house calls."

But a half hour later Phil was at the apartment. He poked me all over and listened to my chest as I breathed in and out.

"Larry," he announced, "I think you have pneumonia. I'm putting you back in the hospital."

Next thing I know, I'm in a wheelchair and being pushed to an ambulance with flashing lights parked in front of my building.

And as the ambulance was speeding downtown to NYU, I can tell you with great conviction that losing the World Trade Center bid was the last thing on my anxious mind.

IT TOOK ANOTHER LONG, disquieting eight days at NYU before I was once again back on the road to a complete recovery. It seemed that when I had been laid up at the uptown hospital, I had caught an infection that turned into pneumonia; that can frequently happen, I was told, when you're too sedentary. At NYU, I was given a regimen of antibiotics and plenty of physical therapy. Olga Kalandova was my therapist, and she was terrific (twenty-three years later, in fact, I still regularly work out with her). She really put me through my paces, first using a walker and then crutches. And it began to work; I was gaining mobility. I wasn't going very fast, but I was learning how to get around. By the time I was released, the pain coursing through my battered body had started to diminish, too. My life, I felt, was on its way to returning to normal. Or at least to what it had been before I had gone off in pursuit of the Twin Towers.

Only no sooner than I get back to the apartment, just as I'm settling in, there's a phone call. It's Steve Roth, the CEO of Vornado. And the winner of the prize I had so coveted.

"I heard what you went through, Larry," he began. "I feel terrible, just terrible for you."

"It was a rough couple of weeks, but I'm feeling much better now. Thanks for calling," I replied. I was genuinely touched that he'd called.

But my health was apparently not the only thing Steve had on his mind. Before hanging up, he asked, "Can I come over? I'd like to get your advice."

"Of course. I'll be glad to see you."

But I was also wondering, *What does Steve want? What sort of advice is he looking for?* He won, after all.

OR DID HE? AN hour or so later I sat across from Steve in my living room, and he revealed that he was having a bout of buyer's remorse. The problem was the Port Authority.

Steve complained that the Port was willing to take his money, but they weren't willing to walk away. They wanted significant control of the complex going forward. They insisted on continuing to make decisions about how things were done in the buildings. Steve felt, and with good reason, that when you hand over $3.25 billion, nobody should be able to tell you what to do.

I did my best to assuage his concerns. I told him that, yes, the Port could certainly be imperious; they liked to do things their way—and only their way. But in my twenty years of doing business with them on Seven World Trade, I had learned that you can work with them. You just need to develop a thick skin. Sure, their demands were often outrageous, but that was the price you had to pay to be in business with a government entity. Besides, once things were up and running successfully, the Port would usually back off. That had been my experience.

That seemed to boost Steve's spirits. But I was not done. I also had something more tangible than my advice to offer him. "Steve," I began to his obvious surprise, "I got this fantastic financing from GMAC. Since I won't be using it, you might as well. I'll call them to set it up."

Steve was immediately enthusiastic. "That's terrific, Larry. Thank you."

Why did I do that? I figured it was the right thing to do. My loss would be his gain. And when Steve left my apartment that afternoon he seemed in a more optimistic mood than when he arrived. He indicated to me that he was confident the deal would close.

BUT I WAS SOON hearing things, intimations that Steve's road to a signed contract continued to be rocky.

My Stroock lawyers passed on the news that they were getting frequent calls from Vornado's legal team. Apparently, Steve's lawyers were not merely troubled by some of the demands the Port was making, they were completely perplexed. And annoyed. To their way of thinking, the Port's positions didn't make sense.

One particularly thorny issue, for example, was the length of the leasehold. Vornado, because it was a public company that used GAAP (generally accepted accounting principles) in their annual reports and to determine the credit rating that allowed them to borrow money at competitive rates, didn't want to carry a $3.25 billion debt on their books for ninety-nine years. Their plan was to divide the leasehold into two periods, one for forty-nine years and the other for fifty. The money going to the Port would be identical as in the single, longer-term lease; however, this bit of juggling would allow Vornado to balance their books and maintain their credit rating. To Vornado's way of looking at things, it was a small accommodation in a multibillion-dollar transaction. Only it wasn't a minor issue to the Port. They wouldn't compromise.

So what do the Vornado lawyers do? On the day before the deadline for the final contract, they sent the Port a signed document— with twelve issues redrafted to reflect Vornado's perspective. It was just twelve issues—out of a contract about as thick as a phone book. Steve's feeling was that these were inconsequential matters, items that could be worked out down the road. He was, after all, accepting everything else the Port had demanded, and he was still paying the agreed price. The Port, however, was livid. They immediately went to their board and began a discussion about terminating Vornado's bid.

It was at that point that I got a call from the same J.P. Morgan banker who had reached me on the day I had been released from the Hospital for Special Surgery. Only now his tone was different.

"Are you still interested in the Towers?" he began cautiously.

"Why are you asking?" I wasn't going to make this easy for him.

"Negotiations haven't gone well. Vornado had twenty-one days to close and they've nearly used up twenty of them. The feeling here is that this deal is not going to get made."

"So?" I asked. I was deliberately being obtuse. I wanted him to spell it all out.

"You were the second-highest bid. We're now coming to you. Are you interested?"

Of course I was interested, but I wasn't going to let him know that. At least not yet. I didn't want to seem too eager and lose whatever leverage I had in future negotiations with the Port; their embarrassment over the failure to close a very publicly heralded deal could, I hoped, make them more compliant, more eager to close. I also was concerned that the Port might simply be using me as a stalking horse, to help persuade Steve that he'd better agree to their terms or else they would move on to someone who would gladly take his place.

"I want to verify that Steve is really out," I told the banker. "Let me speak with him and then I'll get back to you."

So I called Steve, and he confirmed everything the banker had said—and more. "Larry," he stated plainly, "this deal is not going to work. For one thing, Vornado is a public company. They're asking us to do things we just can't do. And for another, it's something more fundamental. They don't trust me, and I don't trust them.

"You want it, Larry," he concluded, "it's yours."

I wanted it all right. And it wasn't long before I was on the phone to J.P. Morgan to tell them just that: I was willing to pay $3.2 billion for the ninety-nine-year World Trade Center leasehold. All we had to do now was negotiate and then sign a final contract.

Later that day, the J.P. Morgan team made it clear that the same stringent rules that had applied to Vornado would now apply to Silverstein Properties. We had twenty-one business days (weekends and holidays stretched the actual time to a month) to consummate the transaction. At the end of this period, there could not be any open issues; all the documents that would be negotiated would be the final binding documents. And on the twenty-first day, along with the signed documents, the Port expected delivery of either a bank certified check or a letter of credit for $100 million. Did I accept those terms?

Yes, I agreed.

And just like that I was, to my colossal surprise, back in the chase for the prize I had so long coveted. The prize that I had been convinced had slipped away. Only now I had another chance.

And once again the clock began ticking.

SIX

I T WAS A WAR party. Gathered in my office on March 20 were my
key people, members of the Stroock legal team, and representatives from our financier, GMAC. All of us were well aware of
the steep hill we had to climb, and how little time we had to reach
the summit. We needed to get our multibillion-dollar financing
with GMAC, our business arrangement for the management of the
three hundred or so stores with our partner, Westfield, and our net
lease with the Port Authority for the entire World Trade Center complex all completely ironed out. And reminding us how difficult this
would be were the headline stories in the morning's newspapers: the
Port had announced that it had officially terminated its negotiations
with Vornado.

With so much to get done, and so little time to do it, the obvi-
ous question was, Where should we begin? But I told the group I
had already worked that out. "Why should we reinvent the wheel?"
I asked. "Why should we waste valuable time trying to get to where
Vornado had already gotten in the negotiations? Let's start where the
Port had ended with them."

"Do you really think the Port will share what they'd worked out?" one of the lawyers worried.

"I'm not going to ask the Port," I replied confidently. "I've a better idea."

I CALLED STEVE ROTH and asked him to share the contract he'd spent the last three weeks negotiating with the Port. I assumed he would be willing to cooperate. After all, he was now out of the competition. And we were friends; he'd come to me for advice when the Port was being obstinate and I, only hours after I had been released from the hospital, had been glad to confer.

Steve, however, saw things differently.

"I spent a lot of money negotiating that document," he began.

I quickly cut him off. "Are you asking me to pay your legal fees?" I asked, immediately seeing where his mind was going.

"Wouldn't be a bad idea," Steve said cagily.

Bad idea? Certainly not for him, I thought to myself. He was asking me to reimburse him for the millions of dollars he had paid to his lawyers.

"I gave you my GMAC financing," I countered, careful to keep my voice steady rather than belligerent. "I didn't charge you five cents for that."

"Well," Steve said breezily, "you're a much more generous guy than I am."

"Generous." I guess that was one adjective that could describe our differences, but at that fuming moment I was thinking of several others. Nevertheless, I ended the conversation with a polite demurral. In business, there's no sense in making enemies; you never know when you'll be working with someone in the future.

So I doggedly tried another tack. I went to the Port and asked if they would share the Vornado document.

They refused. Privileged information, they insisted.

Once again I reined in my dismay. Rather than turning combative or stalking off in a huff, I tried to explain the situation reasonably. "If I can't get this done by the deadline, sure, it's a loss for me," I conceded. "But it's also a big loss for the Port. You'd be throwing away a three-point-two-billion-dollar deal. A deal that the governor wanted. And it'd be the second time that you failed to close the deal." I could have underlined that point by saying something about how embarrassing another well-publicized failure would be, but I decided that would be unnecessary; bad press is something people in the public sector grasp immediately.

And it worked. My low-key argument, with its tacit hint of uncomplimentary media coverage, was persuasive. The Port quickly came back with a compromise: If we give you the documents we have already negotiated with Vornado, we don't expect you to renegotiate those points. You'll need to accept them.

I figured Steve was a shrewd businessman; what he and his lawyers had managed to claw out of the Port would work for me, too. And, not least, now I would have a firm starting point from where I could move on to the myriad other issues that remained. Sure, I agreed.

The Port delivered these documents on Thursday, March 22. We now had till the Port's board meeting on April 26 to finalize all the lease agreements and our financing. It would be a hectic month filled with a marathon of virtually nonstop negotiations. And, as it also turned out, nonstop problems.

YET WHILE THERE WERE hundreds of issues that needed to be resolved, ultimately they all were driven by the same bottom-line concern—keeping GMAC happy, or at least happy enough to write the multibillion-dollar check that I would hand to the Port. Our deal was not contingent on our obtaining financing, but I knew as a practical matter that I could never sign an agreement for the Trade Center leasehold unless I could be certain of GMAC's commitment.

I also knew that GMAC would never remain on board unless

they were convinced that they could go out into the financial mar-ketplace and resell their stake to institutional investors. That meant they had to be confident that there was nothing in the deal I made with the Port that would jeopardize the World Trade Center's mak-ing money for decades to come. They had to be certain that I would make money—so that they would make money, too.

Take, for example, the deal we were making with Westfield for the 430,000 square feet of retail space—stores and food kiosks—that they'd operate as the Westfield City Center. The easy—relatively, that is—part of this negotiation was with the Port. Earlier in the process, we had knocked out a reciprocal easement and operating agreement (REOA), a standard document when there are several owners of an equity and a shopping center is being developed on the property.

The other components, however, proved to be more troublesome. I was good friends with Westfield's Frank Lowy, a great admirer of this self-made man's rise from his childhood in a Nazi-controlled Jewish ghetto to the pinnacle of business success as chairman of an Australian-based multibillion-dollar global retail empire. However, I was unprepared for how tough the back-and-forth would be on the finalization of the plans for Westfield Center.

A crucial discussion involved insurance and restoration of damages—two topics that would come back to haunt us with a ven-geance in the future. For example, the Port insisted that if there was a fire, say, on the ninety-eighth floor, the leaseholder would have to restore the property and also continue to pay his full rent during the restoration. Such a catastrophe seemed, at that time at least, an un-likely possibility, and eventually we and Westfield largely acquiesced.

A more problematic issue was the Port's refusal simply to turn over their plant for us—either Silverstein Properties or Westfield—to run. They insisted on being involved in operating all the mechanicals—heating, air conditioning, ventilation—as well as the security that would be involved in maintaining and protecting the retail center. When you put down the kind of money the Port wanted, you don't want to give up control. And both Westfield and I also believed we could do things more efficiently than they could. For a while, I began to have my doubts that we would be able to work everything out.

Once this was finally done, we then had to sell our agreement to GMAC. To understand how complicated—and potentially problematic—this was, just consider the crowd that marched into those meetings. There were lawyers representing GMAC, Westfield, the Goldman family (who'd be putting up more than two-thirds of the $100 million cash collateral deposit the Port had demanded), and, of course, the team representing Silverstein Properties. It was as if representatives from every white-shoe legal firm in New York had come to this shoot-out. And each wanted to have the last lawyerly word.

And no sooner than this was resolved, then something else popped up. When preparing my initial bid, I had discovered Morgan Stanley, the complex's largest tenant with over 10 percent of the entire commercial space, had in the aftermath of the 1993 bombing at the World Trade Center filed a lawsuit against the Port. Their claim alleged that the Port had misrepresented the security in place to ensure the safety of the towers and so they had the right to walk away from their lease, which was for a colossal 1.2 million square feet of space. And what made the possibility of their leaving the buildings so significant to GMAC was that on June 1 their rent was scheduled to increase by $20 million—$20 million of additional annual income that GMAC had counted on when they had initially agreed to back my bid. GMAC now threatened to withdraw their support if that $20 million couldn't be guaranteed.

And where would that leave me? I would be on the hook for a building I couldn't finance.

So I went to the Port. If Morgan Stanley didn't pay their rent increase, I firmly announced, then I should be able to withdraw from my agreement.

No way, they responded, similarly adamant. You're taking the deal as is—or you're not taking it at all.

But I've heard no before, and in business it is often the signal for negotiations to begin in earnest. At the same time, I also knew that if a pact was going to be made, I would need to be flexible. I had to make clear in my own mind what I could accept and what I couldn't, and between both of those parameters, I would need to find my deal.

It was an enormous undertaking, but in the end we reached a compromise: if Morgan Stanley didn't pay their rent increase, then the Port would pay an identical amount to us. This arrangement satisfied GMAC; the rental cash flow that helped to persuade them to finance my purchase would remain as it had been in their original projections. (In fact, as things finally played out, Morgan Stanley withdrew their suit and paid the $20 million increase in June.)

I breathed a sigh of relief—only to discover that another outstanding lawsuit against the Port threatened to bring things toppling down. An insurance company tenant in Tower One contended that the Port had miscalculated the increase in porters' wages from a recent labor agreement, and they sued for a refund. The dollars at stake in this suit were, in the scheme of a deal costing billions, not significant. However, GMAC reasonably worried that if this one tenant's suit succeeded, then the four hundred or so other tenants would also sue and win. And this would wind up being a sizable sum—money coming out of rental income and therefore taking a large bite out of the cash flow upon which GMAC had predicated their deal. Again, GMAC threatened to walk; when you were putting up the kind of money that they would provide, it was common practice to act unilaterally.

There were several other issues, too, that were unique to this deal and that also gave GMAC misgivings. These all stemmed from the fact that at the end of the ninety-nine-year leasehold, Silverstein Properties would be returning the property to the Port. That meant, at least from the Port's way of looking at the transaction, that despite our paying $3.2 billion, they were merely lending us the complex, not selling it. And therefore they wanted continuous control. The Port insisted that it would continue to maintain police and fire jurisdiction over the complex.

More contentious, however, was the $200 million in capital expenditures that the Port insisted we pay. This sum would be used to make necessary improvements to the complex—including asbestos removal—that the Port had deferred for too many years. GMAC had originally agreed to provide this $200 million fund. Only now the Port insisted that even though we would be paying for the improvements, they would still need to approve, in effect, every nail we hammered. GMAC became enraged. They announced that they were reconsidering funding the capital expenditure budget.

Once again I mobilized my lawyers, and they saved the day. A legal opinion was drafted that made it clear that as a "renter" (albeit a $3.2 billion renter) we in fact did not have to go running to the Port every time we improved the complex. And after they reviewed our lawyers' reasoning, the Port agreed, too.

And while I was busy playing whack-a-mole with every new issue, GMAC was having its own problems. It seemed that the rating agencies were looking dismissively at the deal, and if GMAC couldn't persuade the rating agencies to endorse the financial instruments, they would never hand me a check. And without their money, there'd be no deal: I wouldn't be able to purchase the leasehold.

AND WHAT WAS MAKING the bond rating agencies nervous? It was a concern that had been raised during the early days of my initial due diligence, but then had been put on a back burner as we dealt with more immediate and pressing issues. But while I'd quite complacently shoved the PILOT (payment in lieu of taxes) issue aside, it had remained an open question, and New York mayor Rudy Giuliani had suddenly discovered it. The special property tax arrangement that the city had made with the Trade Center back when it had first been constructed gave him precisely what he had been searching for—a way to elbow his way into a transaction that, he now realized, would make headlines.

In mid-March, just as we were racing to get everything in place by the April 26 deadline, it had dawned on the mayor that the World Trade Center was in fact actually heading toward a sale, and while the governor was an active player in the transaction, he had been left sitting on the sidelines as it all unfolded. Which was the last place this very ambitious politician wanted to be. So he came up with a maneuver that would immediately swing the spotlight back on him. And his proposal—"the Giuliani Threat," we anxiously called it— sure did get my attention. In fact, it seemed as if it would very likely destroy any possibility of a deal.

The mayor held a bombastic press conference to announce that if the proposed agreement for the Trade Center was consummated, the city would sue to terminate the favorable real estate tax provision that had been enacted when the complex had first been built, a shaky time when attracting commercial tenants had seemed uncertain. Capriciously, he had decided that if the Trade Center was going to be leased by a private company, then it should no longer receive the city's help at tax time.

And the Giuliani Threat had sharp teeth. If the PILOT program was removed, the complex's real estate taxes would quadruple, from $25 million to $100 million annually. It was this additional $75 million annual expenditure that was giving Standard & Poor's, the premier bond credit rating agency, the willies. They feared the potential new taxes, when added to the debt that already had to be serviced in this highly leveraged deal, would overwhelm the income. There just wouldn't be enough money coming in to cover what would be going out each year.

So once again, and with the clock ticking away, the Stroock team went to work. They hunkered down for four days and four nights straight researching this issue. When they emerged, they came up with a detailed opinion that Giuliani and the city did not have the legal right to nullify the PILOT program. The argument was basically this: Since this was a genuine leasehold transaction, not a sale, there would be no transfer of title. Remember, they pointed out as proof, the Port had insisted we couldn't, in effect, hammer a nail into a wall without getting their—the property's actual owner—approval. And since the Port was still in ultimate control of the complex, there would be no sustainable way Giuliani could negate the existing tax agreement.

It was a brilliant reversal, one that turned the millstone the Port had put around our necks into an advantage. Only once the lawyers convinced their own partners they were on solid legal ground (and this was not easy; Stroock, quite rightly, was concerned about the firm's exposure if the arguments were built on shaky reasoning or dubious precedents), they immediately needed to set off on another arduous battle—they had to persuade Standard & Poor's.

Bringing the rating firm around was further complicated by their own legal team. They were represented by a Philadelphia firm that, as the Stroock team complained, probably didn't know as much about how real property taxes were assessed in New York City as they did about such matters in their hometown. My lawyers had to educate the out-of-town firm ("babes in the woods," they derisively sneered) about how things worked in the Big Apple and what could and what couldn't be legally aggregated by the city.

Yet even after Standard & Poor's was no longer terrorized by the Giuliani Threat, I still had to work things out with the Port. So I went to them and argued that if I had to face a potential spread of between $25 to $100 million a year for the complex's taxes, it would be impossible to come up with a realistic bid. Neither I nor any future bidder, for that matter, would be able to do the necessary math. Without a firm grasp of the economics of the situation, without knowing the precise real estate taxes, you couldn't calculate the income (or loss) in the transaction. No one would step up to make a $3 billion deal if it had to be based on guesswork.

The Port finally agreed to an indemnification. In the unlikely event that Giuliani somehow did get his way and the PILOT agreement was overturned, they would pay the uptick. That is, if the taxes ever jumped from the present $25 million annually to the threatened

$100 million, then the Port would each year write a check for the $75 million difference.

This was a big victory, and I was beginning to think that maybe I would be able to close this deal before the deadline expired. Except now the rating agencies suddenly raised an issue that left me completely blindsided: they were concerned about the Port's AA rating. How good is their indemnification? they wondered. The Port agreed to cover a $75 million shortfall in taxes, but how do we know, the rating agencies challenged, that they'll have the funds to make good on their heady promise?

Scrambling, I proposed that the Port put this potential payment in escrow. And this (understandably, I concede, but I kept that to myself) left them furious. "How can you think that the states of New York and New Jersey would not stand behind a contract they'd signed? That we'd not be good for the money?" the Port's negotiators raged. I could see their point, but since it wasn't me they had to convince but the rating agencies, I tried another tactic. "Suppose you bond this indemnification?" I suggested. "Guarantee the money that way." This only seemed to fuel their anger.

In the end, it was the lawyers who brought the rating agencies around. They spent days poring over the Port's financial statements, and they were finally able to substantiate to the agencies' satisfaction that the income and assets were sufficient to cover all the shortfalls the Port had agreed to indemnify. Yet while the case was ultimately successfully made by the Port's sharing their books, I also believe it was reinforced by a remark made during this process by one of the Port's officers. "If we need extra money, we just add twenty-five cents to our tolls and we get all the money we could want," he explained with a disarming candor.

And with that resolved, we were nearing the finish line. Except time was running out, and there was still another piece of the deal to put in place. I needed to deliver a $100 million letter of credit to the Port by April 26 or they would walk, canceling the deal. To accomplish this, I had to get about $65 million from the Goldman family investors—but they weren't going to hand over the money until they were satisfied with the finalized GMAC commitment. And GMAC was hanging tough on several of their deal points. The last forty-eight hours—from April 24 to 26—were a nonstop marathon of negotiations. And all the while the Port Authority board meeting at ten a.m. on April 26 was growing closer and closer.

YET ON THE FINAL morning things started to fall into place—as long as the Port directors cooperated by keeping busy with other business before getting to our leasehold contract later in the afternoon. My team conferring with GMAC finally reached a deal just as the Port's meeting had started. They immediately telephoned the Goldman group, which then agreed to release their funds. And I released the approximately $35 million my group was putting up. Only it was now just after one p.m., and the bankers needed to get the necessary signatures to execute the $100 million letter of credit.

When this was finally done, we still had to get the financial document downtown, making our way through the usual bumper-to-bumper New York traffic, and hand-deliver it to the Port board meeting on the seventeenth floor of One World Trade Center. And the clock continued ticking. It was well past two and the board meeting had begun four hours earlier. After all these arduous months, I worried the deal would be lost because of a traffic jam. The check, I feared, would not arrive until after the meeting was adjourned.

Before the board meeting, however, I'd arranged for a sympathetic member to filibuster. He would initiate a discussion about the maintenance contracts at JFK Airport. And it worked—for a while. He succeeded in keeping the board members occupied with these details for several tedious hours. However, as three p.m. approached, he had run out of things to say about the airport contracts, and the exhausted board was at last ready to move on to the finalization of the leasehold agreement with Silverstein Properties for the Trade

Center. Either they would have the money, or they wouldn't. Either there would be a deal, or there wouldn't.

It was at that tense moment that my frenzied banker, after waiting and waiting for what had seemed like an eternity for an elevator in the complex's lobby, burst into the meeting. With a triumphant smile, he interrupted the meeting to hand over the $100 million.

The deal, for all practical purposes, had been made.

EXACTLY THREE MONTHS LATER, on July 24, two days after signing our closing documents with GMAC, I was standing in front of the World Trade Center complex. As news photographers and television cameras recorded the jubilant moment, Governor Pataki handed me a key ring holding an oversized set of cardboard keys (thank God, or else I couldn't have lifted them)—the symbolic keys to the front doors of my hard-fought-for purchase.

At that elated moment, I couldn't help thinking that I had finally realized what had once seemed like an impossible dream. I now

Larry Silverstein holds the ceremonial cardboard keys to the
World Trade Center on July 24, 2001.

owned, for the next ninety-nine years, that is, the premier office complex in the world. I had engineered a deal that would, I estimated, ultimately earn me billions of dollars. And, perhaps most satisfying of all, I now controlled the piece of property that was the commercial heart and soul of the city I loved.

I remember telling Klara, "This is as good as it gets. I've now done everything I've ever wanted to do—and more."

SEVEN

And then, just seven short weeks later, the world turned on its axis. On a clear blue September morning, the first plane, flying at 429 miles per hour, slammed into the North Tower at 8:46; the second plane struck the South Tower at 9:03. Eleven thousand gallons of jet fuel was a savagely effective propellant, and the buildings burned ferociously; in their desperation, people trapped inside began to jump to their deaths. The flames roared out of control for about an hour (a mere 102 minutes, I later learned, was the official time span). And suddenly there was a terrible sound, a great and widening noise that would, in time, echo throughout America and across the world, and the buildings, one after the other, came tumbling down. The cascade of concrete and steel was horrific. In seeming moments, as a thick, toxic gray soot swirled like a plague across Lower Manhattan, the two towers were turned into debris. They were now a grim tomb for 2,753 innocent victims.

The flames and falling rubble from the North Tower had landed with a malicious accuracy on Tower Seven, and the struggle to save this building soon proved impossible. At 5:21 p.m., it, too, collapsed.

And with its destruction, the end of the world as I had previously known it was complete.

Yet in every catastrophe, there are miracles that are the exceptions to the overpowering tragedy. And on 9/11, even on this cruelest of days, there were events for which I will always be deeply thankful. My son, Roger, just as he was pulling up to park in front of Tower Seven that morning, saw the smoke starting to plume from the North Tower and he wisely decided to head instead to our office uptown. And my daughter Lisa happened to leave her home in Westchester later than usual that morning, and so she was stuck in traffic on the West Side Highway rather than at her desk in Tower Seven when the planes hit. I vividly remember repeatedly calling their cell phones, only to discover it was impossible to reach them: there was little cell service in Manhattan that morning. As I watched the events unfold on television my anxiety grew and grew. It wasn't until I saw them walk dazed and bewildered into my office around noon (getting around the city that day was a challenge) that I could consider myself, even on this most wretched of days, blessed. And, as I have already mentioned, I, too, had my own miraculous escape. Thanks to Klara's urging, at the last moment I'd decided not to go to the 107th floor of the North Tower and the Windows on the World restaurant for breakfast but rather to the dermatologist.

As for the rest of the day, what can be said? To know that four of my employees had died, leaving behind six children, was a gaping wound that could not be healed. All the other deaths that fateful day—nearly three thousand people died, a colossal community of heartbreak—had also left me reeling. And so it was that in the rush of overpowering events, the loss of my hard-won prizes, the two towers I had so long coveted as well as Tower Seven, the building I'd constructed, seemed minor occurrences, merely collateral damage in a pernicious attack on the nation. Nevertheless, throughout that devastating day, I found my mind from time to time drifting back to the warm July morning just mere weeks ago when I had stood posing for photographers with those large cardboard keys—keys to a kingdom that no longer existed.

Klara came with me to the office and stood by my side throughout the entire grim day. I remember the phones ringing constantly, and anxiously wondering what each call would bring. Would it be the searing news that one of my employees had perished? Or would it be the joyful announcement that someone had made it out alive?

I remember the elation I felt when, against all odds, Geoff Wharton walked into my office. I had recently hired Geoff to lead the World Trade Center repositioning for Silverstein Properties, and he had been at a breakfast meeting at Windows on the World that morning. But he explained, still seemingly in shock, that he had left to greet a prospective tenant on the plaza and rode the last elevator down before the first plane hit.

Lisa would tell me that I could barely speak; her questions went unanswered, and to this day I have no idea if her words even penetrated my fog of perplexed and troubled thoughts.

As it grew later, Klara took one look at me and immediately took control. "You need to get something to eat," she said with great kindness. Food, she knew, has always been an integral part of the Jewish mourning process.

Along with Geoff and his wife, we went to Primola, an Italian restaurant in the East Sixties, the same one we've gone to for years. We didn't need to order; the maître d' and the waiters just kept bringing dishes to our table as if determined to comfort us, eager to help assuage our grief. Throughout the solemn evening diners came by our table, people I knew and strangers, too, all offering their condolences and, more often than not, their outrage; New York can be a very small town, and we all had lost something on September 11. I don't remember eating a single bite.

When we got back to our apartment on Park Avenue that night, the smell of the destruction all the way at the other end of Manhattan was still strong in the air. A vile, noxious smell—and inescapable.

The immediate aftermath of the worst terrorist attacks in American history

Sleep that night was impossible. I stared out into the darkness for vacant hours, overwhelmed by events that a mere twenty-four hours earlier would have been unimaginable.

AND WITH THE NEW day came the phone calls. It was as if the phone never stopped ringing that week; there were so many friends offering advice or sympathy. But of all the conversations I had, I can still vividly remember three. Two were calls that came unexpectedly, and the third was one that I, looking for a way through my despair, initiated.

The first, as I have already recounted, was an early morning call from Governor Pataki. And what continues to amaze me about our conversation was my instinctive reaction to his opening question: "What do you think we should do?"

I had not, at least not consciously, given this colossal challenge any previous thought. I had been too overwhelmed. And yet once he asked, I didn't hesitate. My response didn't falter. The words poured out with emotion and conviction.

"I don't think we have any options," I began firmly. "We need to rebuild. There's not a doubt in my mind. We cannot allow the terrorists to win. It's not just an attack on the Trade Center. It's an attack on America, on our values. On everything we stand for. We can't let the site remain a pit. We have to rebuild, or else they will have won. I am totally committed to getting this done."

The governor considered this for a moment, and at last he asked, "What would you construct? Would you build the Twin Towers back again?"

Once more, he'd posed a question that I had not previously considered. The events were still too raw. Thoughts about finding a way forward, a strategy for the future, had not entered my mind. Nevertheless, again without previous reflection, as if on instinct, I offered up the outlines of a pragmatic plan. A veteran real estate developer's sensible way to proceed.

"I don't think we rebuild the same two towers," I said. "There'd be a stigma attached to them. It'd be smarter to build something different."

"What?" the governor challenged.

"Maybe four towers, two and a half million feet each," I said without hesitation. "Sixty-, seventy-, eighty-story towers, something like that." I had instinctively done the math and come up with a rough blueprint for replicating the approximate commercial footage that had been destroyed in the attacks.

And with those two spontaneous, yet deeply felt replies—my ardent belief that we must rebuild and an outline for a master plan of sorts for the new construction—I had articulated the two cornerstones that would both anchor and energize my activities for the twenty years that followed.

THE NEXT CALL WAS one I made to a dear friend. I wanted his advice, and his help.

Just four days before the attack on America, on Friday, September 7, the start of a calm late summer weekend, Klara and I had docked our boat in Sag Harbor and then gone to dinner at the East Hampton home of one of my oldest friends, Herb Wachtell. Herb and I, as I have already explained, go back a ways. We had met in 1944 as freshmen at New York's High School of Music and Art—the two of us were piano players. After graduation, we both went to New York University, and when I ran for and won the office of class president, Herb was my campaign manager. And we both had studied law, only Herb had become the founder and senior partner of Wachtell, Lipton, the world's preeminent mergers and acquisitions law firm, and I wound up in the real estate business, heading Silverstein Properties. Also, not least, we both shared the sort of pride of accomplishment that only the self-made can appreciate. We both knew how uphill our journey had been, and how much hard work it had taken to get us there.

The dinner that night at Herb's country home was a sort of special celebration. It had been quite a year for me. I had survived being run over, and although walking was still difficult, I had made my way through a painful injury that could have left me dead or crippled. And if that wasn't sufficient reason to thank my lucky stars, just weeks earlier I had emerged the dark-horse winner of the ninety-nine-year lease on the jewel in New York's commercial real estate crown, the World Trade Center complex. I had won the $3.2 billion brass ring. And who better to enjoy this heady victory with than my oldest friend?

So it was a very enjoyable evening. And it was only later that I would recall Herb's telling me what he had been up to during the past week. He had been in Albany, the capital, appearing on behalf of a client at the New York State Court of Appeals. The case centered on the number of damage occurrences covered by a single insurance policy. Herb explained to me that he was probably the world's leading authority in this arcane field of insurance law. At the time, I simply nodded politely. In truth, I was not that interested; why should

I care about the number of events covered in an insurance policy? Only later would his unique expertise become cruelly relevant to me.

And, to be honest, I still not had made the connection between Herb's appearance at the Court of Appeals and my predicament when I finally managed to gather my careening thoughts and called him on September 12. All I knew was that I needed advice, and I was reaching out to a trusted friend and accomplished attorney.

"I have a problem," I began cautiously once Herb picked up the phone.

"No," he immediately corrected, "you have a tremendous problem."

The conversation went depressingly downhill from there, but it ended with my scheduling a meeting for the next day with Herb and his partner, Marty Lipton, whom I knew and respected from our years serving together on the NYU board.

I met them at their Midtown office and what they said took me by complete surprise.

"You've got to protect yourself," Herb advised. "Unless you get some protection, you'll spend the rest of your life defending yourself from lawsuits connected to the Trade Center. From survivors and from the estates of victims. From anyone who suffered any damage because of the attacks."

"But I only owned the towers for six weeks," I protested. "What can I do?" I asked with mounting desperation.

"You need to get the same kind of guarantees the airlines will be trying to get from the federal government. They'll want protection from wrongful death lawsuits arising out of the planes the terrorists crashed—liability parity legislation."

That was the first time I ever heard about that sort of federal law. But I knew enough to ask, "How do I go about getting it?"

It was explained that it would not be easy. It was retroactive protection that could only be granted by Congress.

My mind was spinning. How was I going to convince Congress to pass a bill designed to protect me? Yet if I didn't succeed, I would be in court for the rest of my days. And if the courts ruled against me, well, I would go broke.

At that disturbing moment I knew that if I were to forge ahead, I would need the best possible lawyers by my side, battling for me. I asked Marty and Herb if they would represent me.

They couldn't say yes, the two men patiently explained. They would need to talk to their partners; it would have to be a commitment the firm agreed to as a whole. The involvement in the rebuilding of the Trade Center, they explained with a prescience I hadn't yet acquired, would demand a great deal of Wachtell, Lipton's resources. The practical problem was whether the firm could focus on what would be an all-encompassing matter, to the detriment of other new business. Would it be an economically sound venture to put so many of their lawyers on a case involving just one client? On a single case that would require so much of the firm's time for so many years? Before taking on such a demanding long-term obligation, they would need the support of their colleagues, they explained.

I left their office feeling worse than when I'd entered. I had been informed that an ominous legal cloud might very well hover over the rest of my life. That I might very well be facing financial ruin. And that the lawyers I had turned to, well, they weren't sure they could represent me.

Then Herb called me later that evening. "Larry," he said, "you're my closest friend. I'm not going to let you down. And the firm is not going to let New York down. We have a civic obligation to help you rebuild from the damage the terrorists caused."

I hung up the phone feeling that I was now starting from a solid place. I knew where I had to begin. If I was ever going to rebuild the Trade Center, I'd need to be protected from wrongful death litigation. This would be my first step. And I was not alone in the looming battle. I had lawyers—and friends—I could count on. Bolstered by their counsel, I would find a path forward. I would not give up. I would not surrender. I would fight on.

· · ·

THE THIRD CALL TOOK me by surprise. It was totally unexpected, and perhaps that explains in part why I was so deeply moved.

My office phone rang on the evening after the attacks, and when I answered a familiar husky, heavily accented voice said, "Larry."

At once I realized I was talking to Ariel Sharon, the prime minister of Israel.

"Arik?" I said, amazed. I was deeply involved with the State of Israel and had raised a good deal of money for the country in my years as chairman of the United Jewish Appeal–Federation of Jewish Philanthropies, but I had not expected a call from the prime minister.

"Larry, how are you?" he asked.

"Arik, I don't know," I answered candidly. "I honestly don't know. I can't figure out what's going on. I'm just beside myself. Everything seems to have come apart here."

"We're worried about you," the prime minister said. His tone was both mournful and consoling. "We are concerned about you."

"Thank you. Thank you, Arik." I was deeply moved. Then a thought occurred to me. "Arik, it must be midnight in Israel. Are you in the office or on the farm?"

"I'm on the farm."

"What the hell!" I exclaimed. It was now my turn to be concerned about him. "What are you doing talking to me? What with the Intifada, all the problems in your country! For god's sake, what're you doing spending time talking to me? You should be asleep."

"You know what, Larry? You've always been there for us. And now it's our turn to be there for you. Because we're concerned about you."

"Now you're making me emotional," I said as I struggled to control the sudden quiver in my voice. "Let me make a deal with you," I finally said. "I'll take care of myself, but you need to take care of yourself. You've a country to run. It's late. You need your sleep. Now get the hell off the phone and get some rest."

"Please take care of yourself," the prime minister reiterated, and then he hung up.

My heart pounding, I, too, replaced the receiver. But not before fully realizing for the first time how what had happened in Lower Manhattan had affected the entire world. And this, too, worked in its indelible way to reinforce my commitment to rebuild.

IN TRUTH, I'M A control freak. I need to be in command, to be able to make the decisions that will affect my family and my business. During the terrible days and weeks after 9/11, I was not myself. I was not in control. For the first time in my life I felt that I was unable to handle things. I felt as if I was a bystander, someone looking on passively as events played out dramatically all around me. I was still a victim being carelessly tossed about by forces beyond my control. And I could see no way in which I would regain the firm hold I had previously had on my life. I was suffering, I later understood, from a form of post-traumatic stress syndrome. And I needed help.

For years I had been involved with the United Jewish Appeal–Federation of Jewish Philanthropy of New York, both as chairman for three years and as a donor. I had worked with them to raise millions of dollars. And now in my time of need, they came to help me.

The federation's Board of Family and Children's Service sent grief counselors to my office. These trained professionals had sessions with me, with Klara, and with many of my employees. The counseling went on for days. They got us all to talk about what was on our minds. The things we had been afraid to say out loud.

It was invaluable. I doubt I could have found the strength to move forward without being able to talk about my concerns, my fears, my doubts with these caring professionals. I would have been lost without the UJA–Federation of New York's assistance.

.   .   .

YET EVEN IN THOSE first raw and inchoate days, I knew that there was one responsibility I had to acknowledge before I could even begin to think about proceeding. I needed to get permission to head into such a consuming undertaking from the one person whose support I most counted on. And to do that, I needed Klara to release me from the pledge I had previously made.

"You know," I began as I sat down next to Klara one evening before dinner, "I had promised you that once we had the Twin Towers, we'd go off and build another boat. Spend some time in the Mediterranean. Cruise around. See places we haven't seen before."

Klara listened in stony silence, as I, for once unable to read her mood, nervously went on. "I know you always wanted to travel. And for me it was always work, work, work. And I know I had told you we would visit all the places you had always wanted to explore once the deal on the World Trade Center was concluded.

"And I will keep my word. I will do that if you want. We'll build another boat and travel the world.

"But," I added as Klara continued to listen without interruption, "I can't travel *and* rebuild the Trade Center. And I can't do any of it alone. So you tell me what you want me to do, and I'll do it."

Klara looked at me with a deep intensity. When she finally spoke, her words were not spoken with disappointment but rather, I felt, with understanding and love. "You're not going to be happy doing anything else—so let's get on with the rebuilding of the Trade Center."

"Sweetheart," I said, bursting with delight, "you're a fantastic partner. Let's get to work."

PART II

INTO THE FRAY

EIGHT

WHEN I'D TAKEN CONTROL of the World Trade Center complex, of course it had occurred to me that some sort of accident could happen on the property. I knew that if there was, say, a fire in one of the offices or someone got his hand caught in an elevator door, I could be sued. And so just like any prudent homeowner who protects himself in case a guest suffers an unfortunate stumble, I'd made sure to purchase liability insurance. In fact, I had a $1 billion policy.

And I had wanted more. However, the Port, in its typically controlling way, had insisted that only certain insurance companies would be acceptable, and after I'd conferred with all the underwriters on their list, $1 billion of protection was the most I could get. Still, I wasn't too worried. A billion dollars is a lot of money. What could happen that'd leave me with a potential exposure greater than such a colossal sum?

On 9/11 I found out.

Therefore, just after I'd finished my conversation with Pataki, I was glad that Bill Berkley, my good friend who was head of W. R. Berkley, one of the major commercial and property insurance companies in the nation, called.

Bill and I go back more than twenty years. I had met him because I was caught trespassing. What had happened was this: When our boat was moored near Key Largo, I would often go jogging or biking through the Ocean Reef Club. And along my usual route I had seen a house under construction that caught my eye. The design was both inventive and beautiful; I'd look at it with a professional's admiration. So from time to time, I admit, I'd ignore the "No Trespassing" sign and go into the construction site to look around.

Only one morning I found myself confronting the surprised owner's wife in the kitchen. "What are you doing here?" she asked anxiously. "Who gave you permission to be here?"

I could quickly see things from her point of view. Not only was she confronting someone who had no right to be in her home, but the trespasser looked, well, rather suspicious. After all, I was in my sneakers and jogging clothes and I had worked up quite a sweat in the course of doing my miles.

I did my best to assuage her fears. I really wanted her to know how much I admired the design of her house. And I explained my professional interest. Then, to make amends, I invited her and her husband to come for drinks and dinner that night on my boat, which was anchored in the harbor.

Marge and Bill Berkley have been our good friends ever since.

And now the morning after 9/11, Bill was on the phone. "Take out a pad," he ordered after only a few brief words of commiseration. "I'm going to give you a list of things you need to do today. The first is the people you need to hire as insurance consultants. And you better hire them fast because if you don't, someone else on the other side will."

I wrote down the names, and when I later called I discovered that several of the consultants he had suggested were already representing other parties. But I did finally manage to hire someone Bill had recommended.

And the next morning, bright and early, Bill called back. "Take out your pad," he ordered again. "Here's what you need to do today."

And that's how it went for months on end. Bill calling first thing every morning with his wise advice. He gave me an expert's battle plan for how I would need to proceed to collect the insurance money I was owed. His advice was invaluable. And what did Bill ask for in return? Nothing. Friends like that are rare, and a true blessing.

Meanwhile, my anxious lawyers were also calling. They warned that the $1 billion of insurance I had taken out would be "totally inadequate."

My only hope, they reiterated, was to get Congress to pass a bill that'd give me the same protection the federal government was already planning to give the airlines: I couldn't be sued for any liability damages arising out of a terrorist attack on America. If I didn't succeed, they again emphasized, my plans for rebuilding the complex would be derailed before I'd even started.

And there's was no way I would let that happen.

"Tell me what I have to do, and I'll do it," I told the lawyers. "Who do I talk to?"

Their instructions were succinct: "You have to go to the leadership of the House and the Senate. You need to talk to every one of them. You need to see them face-to-face and make your pitch for the rebuilding. You need to get them on your side—and New York's."

I WASN'T GOING TO head down to Washington without first making sure I had some allies in the battles that would lie ahead, so I reached out to New York's senators, Hillary Clinton and Chuck Schumer. I had known them both for years.

Back in 2000, when Giuliani was running against Hillary for the Senate, he had called me to say, "Larry, I'd like to give a fundraiser on your boat." So I agreed. I mean, I put up buildings in New York for a living and he was the mayor; how could I refuse? I didn't necessarily like it, but this was a part of how the real estate business worked: you need to buy goodwill.

So Rudy came to the boat. Actually, he didn't even come on board. He stayed on the pier and addressed my forty or so guests, all very accomplished and deep-pocketed people, for just four or five minutes. And all the time, he was talking down to them—literally. He was standing up on the pier, not with them on the deck below. Then there were one or two questions, and he was gone. I didn't like the way it'd worked out, but I figured I had done my part. I had done what the mayor had asked.

The next weekend I happened to be in D.C., at a dinner at the White House. I was in the receiving line, waiting to meet the president and Hillary, when I saw Chuck Schumer approach her and whisper something in her ear. And then she gave me a long look. *Uh-oh*, I said to myself. *That's ominous.*

But when I went to shake her hand, Hillary said, "Chuck tells me that I should ask you to host a fundraiser for me on your boat. Would you?"

The president was standing right there. And now he was looking at me, too. Imploringly.

"Of course," I agreed.

The next morning my first call was to Rudy. I told him what had happened, and then added rather plaintively, "There's no way I could say no."

There was a long silence before Rudy spoke. "Just don't raise more for her than you did for me," he finally said.

So Hillary came to the boat. I mean, unlike Giuliani, she actually came on board. Standing by the gangplank were my captain, the chief engineer, and the chief mate, and she paused to talk to each of them. Then once she was on board, Hillary went up to Klara, and they talked for quite a while. Next I introduced her to other members of my family, and she took her time and spoke with each of them. And the guests, too. It was as if she was really interested in everyone. She even posed for photographs.

When she was done with the small talk and the photos, she gave her speech. She talked for half an hour. A full thirty minutes! And everyone was caught up in her words, following along with interest.

She was preparing to leave when I asked her, "Would you like to stay and join us? We're going on a dinner cruise."

"Love to," Hillary responded.

And we loved having her. From that evening on I was a big fan and big supporter of Hillary's. Oh, and the next morning, first thing, she called Klara to thank her for the wonderful evening.

(As for Giuliani, it wasn't long before our relationship, such as it was, became further frayed. I went to a fundraising breakfast where he was speaking and he made what struck me as disparaging remarks about the LGBTQ+ community. So I went up to him and said, "I have a gay daughter, and I love her. And I found what you said offensive. So I'm leaving." And just like that, I walked out.)

I also had a longtime relationship with Chuck Schumer, New York's senior senator. I had been a contributor to his campaigns when he ran for Congress, and I liked him, too. We had the same sort of values, same sort of life experiences. We understood each other.

But back when there had first been talk that Chuck, a Brooklyn congressman, was contemplating running against Al D'Amato, the incumbent Republican, for the Senate, I had tried to talk him out of it. "D'Amato has very deep relationships," I warned Chuck. "He does favors for everyone. I'm concerned for you if you try to take him on."

But Chuck wouldn't hear it. "I'm going to run against him," he said. "And I'm going to win." He was very confident.

But I still wasn't. And so I found myself at a fundraiser for D'Amato. It was a small group, all significant people in their fields, and there was no press. It was all very private, very cozy. And I guess that was why D'Amato felt he could talk freely. And he sure did. He really tore into Chuck. Called him lots of names, including a "putz-head." I didn't like what he was saying at all. It wasn't just inappropriate, but the derisive tone struck me as, well, anti-Semitic. After that, D'Amato was a nonperson for me. Sure, I contributed to his campaign, and I asked him to help me push some projects forward, but we had no personal relationship. And I grew even closer with Chuck.

And now as I was contemplating going down to Washington, I reached out to New York's two senators, both my friends. I called Hillary's office first, and then Chuck's. Neither was available, but Chuck got back to me pretty quickly.

I gave Chuck my pitch. I told him what I needed Congress to do if I was ever going to get down to the hard job of rebuilding what the terrorists had destroyed.

"You're a very good friend, Larry," he began. "You need this legislation, and you're entitled to it," he went on. He concluded our conversation with a pledge that left me thrilled. "I'm going to get this for you," he promised.

Not much later Hillary called, and I told her about my conversation with Chuck. "If he said he'll do it, you can count on it," she stated. "But if you need my help, I'll be there for you, too."

So now I had New York's two Democratic senators in my corner. Nevertheless, my lawyers advised, I would need more help if I was going to succeed in pushing a bill through a divided Congress. You will need to hire a lobbying firm, I was told. They will be able to get you in to see all the powers that be in the House and the Senate.

Okay, I agreed. But finding a lobbyist turned out to be a lot more complicated than I had imagined. And so I was fortunate that I had two knowledgeable friends who lived in D.C., Gerry and Ellen Sigal. Gerry ran a very successful construction firm, and Ellen was the founder and chairperson of the Friends of Cancer Research, and had also served on the boards of several government cancer research groups. They were able to offer wise insider's advice about how things worked in Washington.

Still, when I started going down the list of possible lobbyists with my lawyers, we had to rule out one firm after another. It seemed they all had clients that would put them in direct conflicts of interest with me and my determination to get Lower Manhattan rebuilt.

Finally, though, we found a firm unencumbered by conflicts— Quinn Gillespie & Associates. Jack Quinn had been Al Gore's chief of staff until President Clinton had poached him to be his counsel, while Ed Gillespie was Mr. Republican, a man with very close ties to the current Bush administration. By hiring them, I figured I would get a foot in both political camps, and that could prove invaluable in the partisan battle that inevitably lay ahead. And so I signed them on, paying the usual monthly fee for their services. However, there was one part of the deal that left me a little unnerved: Quinn Gillespie would get a $1 million bonus if Congress passed the parity legislation I needed. That seemed a lot of money to be paying for something that I should have as a simple right; after all, the entire country had been attacked, not just my buildings. But I decided not to argue. I needed to move forward without delay.

Yet there was also something else I needed to do before going down to Washington. Truth be told, I had not completely recovered from the accident seven months or so earlier. I was still dealing with pain, and despite the ongoing physical therapy sessions, I continued to have difficulty walking; I was often relying on a metal walker for support. I didn't feel I had the physical strength to head down to Washington on my own.

"Sweetheart," I asked, "would you do me a favor? I don't think I'm up to doing all this. Would you come to Washington with me?"

"Of course," Klara agreed without hesitation.

OVER THE NEXT FEW months, Klara and I spent a good deal of time walking through the corridors of power. In fact, the first thing I pain-fully learned about Washington was that it was built on the horizon-tal, nothing taller than twelve stories. It was not like New York, a town of high-rise buildings with elevators that would zip you up to the top in a jiffy. In D.C., especially in the Capitol Building, you had to walk everywhere, rambling down long, marbled corridors that seemed to stretch on forever. It really wore me out. Without Klara's standing by my side trip after trip during those difficult months, I don't think I'd have found the stamina to trek through the vast hallways.

With Chuck Schumer and Quinn Gillespie paving the way, I managed to get meetings with all the key congressional players. You'll only get fifteen minutes, they told me, so don't waste them. Be prepared. And I was. I made sure to polish and then rehearse my pitch. In the real estate business, you learn to be a salesman, but this was different. I would be arguing not just for a piece of business, but for the future of my commercial life. I took the challenge very seriously.

In congressional office after office, I would go into my spiel about why I felt it was very necessary to ensure that valuable years should not be frittered away fighting wrongful death suits. It was vital to get back to the rebuilding at once. The World Trade Center site could not be allowed to remain a gaping hole. And this was important not just to me, or to the Port Authority of New York and New Jersey. The restoration of Lower Manhattan was crucial to the entire nation, to the way the world would view America.

It seemed to be working. Most people seemed to get what I was saying. Many, in fact, asked a lot of good questions, and often the conversations stretched on for an hour or more. Others, though, didn't grasp why the legislation was needed. They weren't at all knowledgeable about the issues involved. They would simply thank me for coming and then an aide would escort me to the door. And I would leave wondering if I was deluding myself, if I'd ever get Congress to pass the bill I needed.

I grew more optimistic, though, after Klara and I met with Trent Lott, the Mississippi Republican who was the powerful Senate majority leader. I had headed into his office figuring that he'd be a tough sell. But after I had finished my presentation, he looked me in the eye for what seemed like an eternity. Then he spoke.

"Anytime you get Chuck Schumer and Trent Lott on the same page, you got yourself a deal. And let me tell you, we're on the same page on this one."

With that earnest pledge ringing in my ears, the trek back down the hallway afterward for once didn't seem so long.

YET A FEW DAYS later, after I was back in New York, I got a disturbing call from Chuck Schumer. "Larry, we got a real problem," he began. The problem was Senator Don Nickles of Oklahoma. "He's being very difficult," Chuck said. "And he can stop us. He can exercise his senatorial prerogative and the parity legislation will never reach the floor for a vote. You have to get him on our side."

"Then I will," I said very confidently.

"You don't understand," Chuck continued, his voice despairing. "He doesn't like New Yorkers. Particularly wealthy New Yorkers. And even more particularly, developers."

But what choice did I have? I had to persuade him. If I didn't succeed in getting this bill, it would end my business life. I was determined not to fail.

Ed Gillespie, my Mr. Republican, set up the meeting, and he accompanied Klara and me to Nickles's office. We sat opposite the senator, and, after very little small talk, I started making my pitch.

Usually my argument—the inherent logic of individuals not being unduly victimized by an attack against the entire country, the need to send an uplifting message to the world—all seemed to strike a chord. People listened with interest and sympathy. When I went to George W. Bush's White House, for example, they wound up keeping me there for over an hour. They didn't want to let me go. But this senator? He just sat there as if his face had been chiseled out of stone. He looked at me with hard, blank eyes. So I finally stopped. I figured I had better hear what was on his mind.

"Mr. Silverstein," he said, "the federal building was destroyed in

Oklahoma City. We lost one hundred and sixty-eight people. But the world didn't come to an end. But when the Twin Towers are hit in New York, the world's supposed to stop?"

"Senator, with all due deference," I respectfully replied, "it was not just an attack on the Twin Towers. It was an attack on America. That's why it's incredibly important for us to rebuild. We can't let our enemies win."

The senator was not moved. "When our building came down in Oklahoma City," he said matter-of-factly, "your people didn't do a thing to help us. And in New York they're raising hundreds of millions of dollars."

Your people? I was livid.

I was prepared for his attitude toward New Yorkers, but I could not help but wonder—perhaps wrongly, I concede—if there were deeper resentments provoking his antagonism. Before going to the meeting I had done my homework. I had called the UJA–Federation in New York and asked them to find out what the Jewish community had donated to the survivors of the Oklahoma City bombing. It didn't take them long to get back to me with the amount—$10 million.

And now I shared this number with the senator.

"Mr. Silverstein, I am not impressed," he declared with an icy disdain.

At that point, my heart started pounding like it was about to jump out of my chest. I could feel my blood pressure rising. I was so angry that I thought I would have a heart attack if I had to spend another moment sitting across from this jerk.

"Thank you for your time," I said. I shook his hand politely, and as quickly as I could, I made my way out of his office. But as I was leaving, someone I hadn't paid much attention to before, a young man who been standing in the back of the office listening quietly the whole time, reached out and gave my hand a surprisingly warm and firm grasp. And he looked me straight in the eye and said, "We're gonna help you." But all I was thinking was, *Yeah, sure.*

As soon as I was in the hall, I turned to Ed Gillespie. "That was a goddamn disaster," I wailed.

"No it wasn't," Ed insisted. "It was a terrific meeting."

"What are you talking about? Didn't you hear the senator?"

"Forget about the senator," Ed declared. "His legislative aide said he's going to help you. It's the aides who have the real power in Washington."

And wouldn't you know it, Ed was right. We wound up getting Nickles's vote. In fact, we got everyone's vote. President Bush signed a bill that not only protected me from wrongful death suits, but also the Port Authority, as well as Massport, the entity that controlled Boston's Logan Airport, where the flights of two of the planes hijacked on September 11 had originated. And at the last minute, Boeing Airlines also received liability protection in the bill.

When I signed the $1 million bonus check for Quinn Gillespie & Associates, I did so without the least bit of dismay. In fact, I felt it was a small price to pay. After all, now I would be able to get on with the real business that I'd been forced to postpone: I could start making plans to rebuild the Trade Center.

YET WHILE I HAD been trying to move things forward in Washington, back in New York the local officials were also busy—busy with their attempt to wrest control of the Trade Center complex from my hands. Just three weeks after September 11, Giuliani, the lame duck Republican mayor of New York, and Pataki, the Republican governor of the state who was running for his third term (while also, according to several press reports, already contemplating a run for the White House), took it upon themselves to form the Lower Manhattan Development Corporation (LMDC). Its purpose? The LMDC would now control and distribute much of the nearly $20 billion in federal funds that, thanks to Chuck Schumer's efforts and persistence in the Senate, had been pledged for the rebuilding of the downtown area. It was a lot of money, and it would create a lot of jobs—and in the process it could win a lot of votes for the local Republicans who controlled this gusher of dollars coming from Washington.

I was never consulted when the LMDC was formed. It didn't matter to these politicians that I had a valid lease for the next ninety-nine years on the sixteen acres of land where the Towers had stood. Or that I was still obligated to pay $10 million a month in ground rent. Or, for that matter, that I had a contract with the Port that not only gave me the sole right to rebuild the buildings that'd been destroyed, it *obligated* me to replace them. Giuliani, though, didn't want to replace the towers. "This is going to be a place that is remembered a hundred and a thousand years from now, like the great battlefields of Europe and the United States," he said in a speech during his last days in office. Instead of commercial buildings, he wanted the site to be used for "a soaring, monumental, beautiful memorial that just draws millions of people here that just want to see it."

And as a first step toward accomplishing that remarkably passive goal, the mayor wanted me to relinquish my right to rebuild. "What would it take for you to walk away?" he asked. "Fifty million dollars?" When I didn't respond, he misread my stunned silence for a negotiating ploy. "Come on, Larry, everyone has a price. How about two hundred and fifty million?"

Giuliani, as I interpreted his offers, was trying to get me to announce a number that would cover the money I had invested in lawyers, accountants, and architects, not to mention my time, and come up with a number that would allow me to walk away with a tidy profit. He wanted me to give him a number that would make me whole. Sure, I had already spent tens of millions of dollars, probably over $75 million at this point, but I refused to do this self-serving math.

Instead, I knew I had better make my position crystal clear. "There's no amount of money that will convince me to walk away. I have a responsibility. To New York City. And to America. We need to show that we were not defeated. That we're not simply mourning. We need to demonstrate that we are moving forward. That New York will remain the commercial capital of the world. And I am determined to do just that."

Yet despite my heartfelt conviction, if I had any doubt about who was running things, I just had to go to the site: it was cordoned off by federal troops. I had to get permission to walk on land where I was paying $10 million a month in rent. I was fuming. It was a disaster area, filled with tons of rubble and the remains of those who had perished when the towers had collapsed; I understood that. Yet I also couldn't help feeling at that tense, uncertain time that the politicians were trying to take something from me that was rightfully mine. More importantly, they were trying to stop me from fulfilling what I felt was a sacred duty: to help New York rise up from the ashes.

Sure, I listened to what the politicians had to say. They were telling me that this sort of project was too big for a private developer. That only government agencies could get the job done. But I knew this wasn't true. The Port hadn't built anything of significance in over four decades. And look at the airports they ran—LaGuardia, JFK. They were monuments to mismanagement. They just didn't have the expertise to lead a project of this size and scope.

What this really was about was money and power. The rebuilding would be a vast undertaking, and that meant a lot of money would be spent and a lot of jobs would be created. The government officials wanted to have their agencies award the construction contracts. They wanted to dole out the cash. If they controlled the purse strings, then they would get the credit for all the construction and contracting jobs—and they'd get the votes. And they would make a lot of rich, powerful friends—all potential donors.

I looked at their arguments and gave them some thought. But in the end I believed I knew precisely what motivated their desire for control: it was a cynical, self-serving power grab. They were out for themselves, not the city.

Yet, at the same time I was sympathetic to the surviving families who had lost loved ones on 9/11. Many of them felt that the World Trade Center site was hallowed ground; it contained, after all, the vaporized remains of those they had lost. Grieving, they raised their anguished voices to demand that the entire site—all sixteen acres—be turned into a memorial park. It should be a monument, a sacred place, rather than a bustling center of commerce. I listened and my heart went out to them. I invited them to come to my office and we sat and talked. I felt their deep pain. But I also believed that the future of New York was not about grieving or sentimentality. It would be about rebuilding. It would be about restoring the beating commercial heart of downtown Manhattan. It would be about showing the terrorists that they had not won.

And so I realized that if the complex was going to be rebuilt, I needed to find a way to assert my control. I had to let the politicians know that I was not relinquishing my legal "right and obligation" to proceed with the reconstruction. And as a practical matter, I needed to convince them that I could get the job done: that I not only could erect a building quickly, but also that it'd be a commercial success, a high-rise office structure that would attract tenants. And just as I was wondering how to accomplish all this, Con Ed, the utility that provides electricity in Manhattan, handed me the perfect opportunity.

"WE NEED TO MAKE sure the lights stay on in Lower Manhattan," a clearly concerned Gene McGrath, the chairman and CEO of Con Ed, called to tell me just weeks after 9/11. "For the time being we're okay using emergency generators, but that's just a stopgap measure. They can't do the job forever. We need to replace the electrical substation that was destroyed."

On September 11, when the North Tower had started to collapse, the 362-foot antenna on its roof broke off. It went hurtling down, slicing like a long, pointed spear through the façade of Tower Seven, the adjacent forty-seven-story building I had erected back in 1987. A massive fire started raging. The sprinklers were useless because there was no power in Lower Manhattan. And although a small army of firefighters battled valiantly, the flames grew too intense. At

Excavating the footprint of Seven World Trade Center
to prepare for the new tower

5:21 that evening, the structure gave way, and the building collapsed. Fortunately, no one had been inside when it came down. Not a single life had been lost, thank God. But the collapse had destroyed the Con Ed substation at the tower's base that produced the electricity for most of downtown, including Wall Street.

"We need to rebuild the substation at once," McGrath told me. "And not only that, we'll need to make it bigger. We'll need space to put in a hell of a lot more transformers than we had before." Then he added, "And if you're not going to do it, we will. We'll build the substation ourselves."

I didn't take this as a threat, but rather as a challenge. And it got me thinking. By the time later that same day when Governor Pataki called me, my mind was made up.

"How quickly can you rebuild the substation?" he asked. Clearly, McGrath had contacted him, too.

"Well, funny that you ask, Governor. I'm not just going to rebuild the substation. I'm going to rebuild the whole building."

"The whole building?" he repeated. He sounded dumbfounded.

"Yes," I said firmly. "And the new substation will be its base."

And just like that, I had articulated my plan and my mind was set. I would piggyback onto the pressing need to get the substation built as quickly as possible. I would plow ahead and get a new Tower Seven erected on Con Ed's accelerated schedule, too. And because the substation was such a necessity, I would have the politicians on my side, maybe even cheering me on; after all, since no one had died in Tower Seven, they wouldn't have to fear critics arguing that this was hallowed ground. The focus would not be on me, but on keeping the lights on in Lower Manhattan. Yet as work proceeded, I would be tacitly establishing that not only did I have the right to build on a site where one of my buildings had stood, but I also could get the job done. If I succeeded with this building, it would be the precedent for everything that would hopefully follow.

BUT WHAT KIND OF building should I put up? Did I really want to build a replica of the old Tower Seven? In the 1980s, I had set out to put up a relatively inexpensive office tower that would attract tenants, and I had done just that. It was a development strategy that had turned Tower Seven into a reliable moneymaker for Silverstein Properties. But architectural critics had sneered that it looked like a "shoebox" and their derision was painful because, well, they had a point. Now that I was older and the passing years had given me a comfortable cushion of success, I wanted to build something that would fill me with a greater pride. I didn't want to use the sort of heavy, even lugubrious red granite that had covered the workman-like façade of the previous tower. In my mind's eye, I envisioned a building with a luminous curtain wall, and expansive floors without columns. I wanted a building that offered its tenants unobstructed views. And I also knew that this tower, the first to go up in the Trade Center complex, should be a bright, shining example of the sort of significant architecture that would help to revitalize a Lower Manhattan still reeling from being Ground Zero in a terrorist attack on America.

I went to David Childs, the inventive architect at Skidmore who I had previously hoped would redesign the lobby of the old Twin Towers, and told him I wanted to hire him to design the new Tower Seven. And he needed to do it as quickly as possible: we had to get the electrical substation up and running.

David, however, immediately slowed me down. "We need first to think about how this new building will affect the entire site," he lectured. "Before we go ahead with our design, we need to begin thinking about a master plan for the entire complex."

Now, I knew that the politicians at the LMDC had begun talking about a master plan. And I imagined they would still be talking for months (and months) to come. But Con Ed couldn't wait around idly until they had drafted a plan. Wall Street, the financial capital of the world, needed a reliable source of electricity to keep its engines churning. I decided that I would make some independent decisions, and then the master plan would have to deal with them.

One of the first things I decided, and David wholeheartedly concurred, was to eliminate the massive superblock that had been fashioned when the Trade Center had been built. It was thirty-five feet above grade (that is, it had been significantly elevated above the preexisting street), and it created an artificial thoroughfare that was nothing at all like the street grid that had existed before the complex had been built. And it just hadn't worked. People hated the superblock. One of the charms of downtown New York was walking through its quirky intersecting streets. But they had been cavalierly demolished. The old Tower Seven, in fact, had blocked Greenwich Street

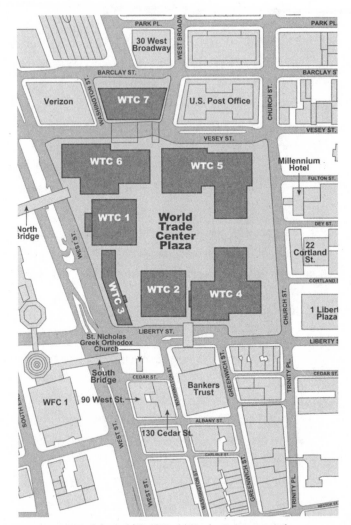

Map of the original World Trade Center complex

from flowing south through the complex, and part of the building had been built directly on top of what once had been a pedestrian thoroughfare.

So David and I came up with a plan to reopen Greenwich Street. We would build facing the street, not directly on top of it. We would restore the street grid to what it had been before the towers had gone up. We would remake the area as a neighborhood rather than simply a foreboding commercial complex. And because I under-

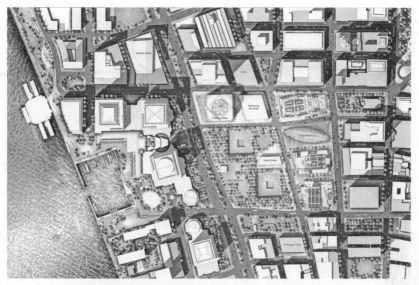

Master plan map of the new World Trade Center, now
connected to the New York City street grid

stood that one day in the future, I would undoubtedly have to deal
with the master plan the politicians finally decided on, I kept both
Governor Pataki and Mike Bloomberg, the new mayor, informed of
our redesign of the street grid. They were supportive. But at the same
time, I didn't wait for their official approval. I kept moving forward.
The only way to get this done quickly, I reasoned, would be to present them with a fait accompli.

Was this rash? Maybe even arrogant? Who was I to take things so
peremptorily into my own hands? But this was a crisis. The Con Ed
substation that supplied electricity to nearly all of Lower Manhattan,
that powered the commercial engines of Wall Street, that helped keep
the very heart of our nation's economy beating, had to be rebuilt as
soon as possible. If I didn't take the lead, it wouldn't get done.

YET DESPITE MY GUNG-HO determination, it wasn't long before I had
to slow down to confront a problem I hadn't previously fully considered. The old building had been a 2-million-square-foot tower.

That was a lot of leasable space. And renting offices was what drove the commercial real estate business. It was how a developer made money. However, when I started working with David on laying out the floors of his newly designed tower, he informed me that since I was re-creating Greenwich Street, I would have a smaller site. And that meant smaller floors. The old floors had been 47,000 square feet each. The new design worked out to be significantly more compact: 40,000-square-foot floors. And that added up to a lot less leasable space.

I was chagrined. I had the legal right to build the same size building as the old Tower Seven, an identical 2 million square feet. Yet I also knew that a contract right was one thing, and a moral duty another. In the aftermath of 9/11, I felt I had a responsibility to correct the mistakes of the past by bringing back the street grid that the original planners of the Twin Towers had eradicated. I wanted the neighborhood to come back to life.

At the same time, I am also a practical businessman, so I told David to come up with a design that would allow me to have both a pedestrian Greenwich Street *and* a 2-million-square-foot building.

He did. He pulled the building back to the east, and the larger floors now cantilevered over Greenwich Street as they rose up. It was an ingenious design. It got me the 2 million square feet of space I was legally entitled to build. Only it looked awful. The architecture was just out of place. In fact, the more I stared at the mock-up, the more I came to the conclusion that the building was as ugly as sin.

But if I scaled back the design to 1.7 million square feet, I would be voluntarily giving up 300,000 buildable feet. Over time this would cost me about $300 million. Maybe, I asked myself, I should forget about re-creating Greenwich Street. So what if it would help bring back life to a battered neighborhood? I was a developer, not a city planner. I argued back and forth with myself. I could make a persuasive case for either design.

But in the end, I knew I had to do the right thing. The first building constructed after the 9/11 attacks would set the tone for the entire complex. It had to look right. It had to feel right. Hell, it had to be magnificent. There was no other way. This meant I had to forget about getting every dime I could out of this project. Something more important than the bottom line was at stake.

I told David, "We're sticking with Greenwich Street. Let's go with the smaller, straight-shaft tower. Let's make it the best building we can."

David designed a masterpiece, a fifty-two-story tower with taut glass floor-to-ceiling walls and stainless-steel spandrels that would reflect the shimmering sunlight. It was beautiful.

I was very proud of what we were setting out to do, and I wanted to make sure the local community board, which had the power to veto the plans, fully understood the intensity of my commitment to the neighborhood. I wanted them to appreciate the care and thought that had gone into the design. So I made the decision that not just David would address the board, but I would, too. It would be a public appearance that, I knew, could backfire. New Yorkers always say what's on their minds, and I was putting myself directly in their line of fire. But I was determined that they understand our vision. And I felt that no one could share it better than me.

Still, it was not, I admit, without some anxiety that I stood up before the community board in May 2002 and spoke. Their criticisms could set the project back months, maybe even years. And do you know what happened when I finished? They applauded.

Now all we had to do was build it.

NINE

A S I PREPARED TO move forward in 2002 with the construction of the first building that would rise near Ground Zero,
life downtown was still very much a work in progress. A
master plan for the future of the Trade Center complex had not been
settled. Police security barricades continued to block off many of the
streets; construction workers had not nearly finished clearing the site
of the one million tons of concrete and steel that had come tumbling
down on September 11. And there was a good deal of heated public
debate over whether commercial buildings, or anything at all for that
matter, should be built where nearly three thousand people had lost
their lives. To many of the victims' families, the ground was sacred,
the resting place for the immortal spirits of their loved ones. But
with Con Ed prodding me on, I was determined to move speedily
forward. I saw an opportunity, and I had made up my mind to take it.

Only as we got ready to break ground for the new Tower Seven, I
encountered a sudden obstacle—one that could delay construction
for years, maybe even prevent it forever. Worse, it was an impediment
that I'd created.

In order to accommodate the reopening of Greenwich Street and still have a well-functioning office tower, I had approved the extension of the building's footprint. It now reached out into Vesey Street. It was only a small extension, just a strip of land that ran adjacent to Greenwich Street. But it was land owned by New York City, not by Silverstein Properties. And my lawyers were adamant: You can't build on land you don't own. It could jeopardize the entire $800 million project.

You didn't have to be a lawyer to see their point. So I met with city officials and explained the situation to them, and they were very understanding. They were willing to trade the required strip of Vesey Street for the reopened Greenwich Street. It would be a win for the city and a win for me. "Great," I agreed. "Let's get it done."

Except the city couldn't get it done, or at least not as quickly as I needed. Before the strip of Vesey Street could be conveyed to Silverstein Properties and the Port Authority, a tall mountain of regulations and environmental laws would need to be addressed. And that lengthy, painstaking process was just for starters. The land could only be acquired through condemnation, and that, too, would be another tedious exercise. It could drag through the courts for years. And if it were challenged, not only would more time—years?—be spent in the courthouse as the process played out, but there was always the possibility that I could lose. I could spend the next decade fighting for this strip of land, and at the end I could walk away with nothing.

There had to be a better, quicker way, I wanted to believe, one that came with a guarantee. So I went to Mayor Bloomberg and explained the situation.

He was sympathetic; he wanted to help. After all, the mayor understood the importance of getting the new electrical substation up and running without delay. And Mike Bloomberg was a shrewd businessman; he knew how to make deals. But in this situation, the mayor explained with regret, his hands were tied. He didn't have the power to overrule the mandated condemnation process. The best he could do, the mayor offered, was to give me what amounted to a

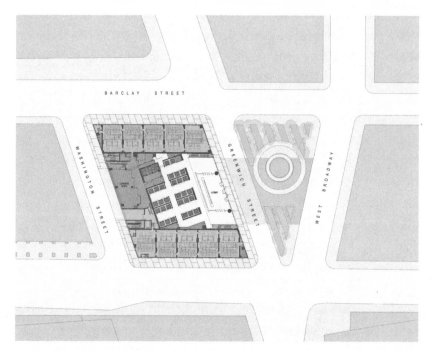

Street-level plan for the new 7 World Trade Center

Architectural rendering of the base of the new 7 World Trade Center

handshake agreement. "Somehow," he assured me, "the required strip of Vesey Street will eventually be transferred to the Port and Silverstein Properties."

"Eventually?" one of my lawyers cried when I relayed the conversation to them. "A handshake agreement?" He barked that it was absurd to proceed on such a flimsy guarantee. "It's not your land. You're putting an eight-hundred-million-dollar investment at risk. You could be tied up in lawsuits forever. And if the mayor fails to deliver, you could lose. Only it's not just eight hundred million dollars. You could be exposing yourself to tens of millions of dollars in additional punitive damages."

I knew they were right, but I wasn't ready to give up. "Do you have any suggestions? Is there a better way to get this done?" I asked, nearly pleading.

The lawyers were stymied. The only prudent course, they patiently explained, was to put off construction until the entire condemnation process had transpired and the Vesey Street strip was officially conveyed by the city. And that, they acknowledged, could take years.

Years? *The hell with that,* I said to myself. *I need to get this done.* This was not the time to be cautious. It was the time to forge ahead. To take an extraordinary gamble: a gamble on the future of New York.

I began driving piles for a fifty-two-story skyscraper into a strip of land I didn't own.

CONSTRUCTION COSTS MONEY. As I started buying the steel that would gird the new tower, I began to worry that I still hadn't received a penny from the companies that had insured the original building. I was due approximately $861 million—and it was money I was counting on to pay off the $489 million mortgage on the old tower and help pay for its replacement.

In the days following the collapse of Tower Seven, I had reached out to Industrial Risk Insurers, a unit of General Electric's Employers Reinsurance division, who had insured the tower and filed my claim. They requested some confirming documents, but this was just a formality, they assured me. Once the papers were received, the $861 million, I was told, would be forthcoming. So my people gathered all the necessary papers and swiftly sent them on. And I started waiting for the check.

Six months later I was still waiting. So I got on the phone. "What's going on here?" I asked. "What's with you guys? What happened?"

We need a little more time. And a little more information, they explained.

"Well, what is it that you specifically need?" I inquired.

It turned out that they needed a lot. My staff would have to assemble cartons and cartons of additional documents. That prospect increased my anxiety. "Another truckload of information!" I exploded. "Are you guys for real? Are you going to honor your obligations?"

They acted as if I had insulted them. Absolutely, they said with wounded indignation. Absolutely. Absolutely. We are men of good faith.

So once again we rounded up everything they were demanding. And after we sent it off, I asked, "How much longer should it take?"

Certainly not more than six months, I was assured.

Six months? Another six months? But what choice did I have but to wait? I needed the money to proceed.

When the six months passed, I reached out to them once more. So? I inquired.

We need more information, they responded.

That's when I finally realized what they were trying to do. The original insurance agreement I had signed had specified that a replacement building must be built no later than two years after the destruction of Tower Seven. If it wasn't erected by then, the insurer's exposure would be significantly limited. And that, I now thought, was their catch-22 strategy: I couldn't build because they hadn't paid, and because I had failed to rebuild they wouldn't have to pay.

I exploded. "You guys are full of you know what!" I bellowed.

"You are no more intent on helping me finance this building than you are on taking a rocket to the moon. I don't believe you for one minute. I'm entitled to payment and you'd better pay up. Because if you don't, well, goddamn it, I'm going to sue the hell out of you."

I don't often lose my temper, but when I do it's a doozy. And this was one of those times. I was livid.

But my anger was effective. The insurers, I now believed, understood I was serious, that I wasn't making an idle threat. They realized I was indeed going to sue. And they were smart enough to grasp that not only would I win, I'd collect sizable damages in addition. They knew they had no case at all.

At last they sat down at the negotiating table and a deal was worked out that would get me the money I needed to start building. But the runaround from the insurers had taught me a lesson that I would need to take to heart throughout the entire process of rebuilding the Trade Center complex: to get paid, I'd have to battle every step of the way. In fact, as things would turn out, collecting the insurance claim on Tower Seven would be the least complicated episode in my ongoing struggles with the insurers. When I took out the policies on my buildings in the Trade Center, I paid enormous premiums. And the companies had been only too glad to pocket the payments, and in return, they professed legitimacy, honesty, and a sense of obligation. I was about to find out it was all a big lie.

DESPITE THE TAIL-DRAGGING BY the insurance companies, in May 2002, we began to build. It was a fast-track schedule, simultaneously moving forward on the Con Ed power substation that would be the first seventy-seven feet of the building, as well as on the office tower that would rise above. We were charging ahead so rapidly, in fact, that definitive construction plans had not been completed. We had to get the power station operating as soon as possible. And the only way to accomplish that was to work out many of the construction details in the course of building.

Yet that sort of freewheeling approach didn't just create an unprecedented construction problem. Proceeding on two fronts without finalized plans and specs led to a unique financial situation, too. In January 2003, the city's Industrial Development Agency had approved up to $400 million in tax-exempt Liberty Bonds for use in rebuilding 7 World Trade Center. But there simply was no way to allocate costs between Con Ed and Silverstein Properties in advance because we truly didn't know how it would all work out each week. There was no way to be certain as we headed into the workweek what specifically we'd be doing, let alone how much would wind up being attributable to the substation or to the office tower. Therefore, there was no reliable way to determine in advance who should be paying for, say, the elevator banks, the foundation, or the sections of the façade that we would be fabricating in the days ahead. Would this be work for Con Ed or for Silverstein Properties? Sure, we had a rough plan of what we would be doing, but we didn't have the sort of very specific blueprints and cost estimates that are usually in place before a building goes up.

Meanwhile, our general contractor, the Tishman Construction crew, also needed to get paid. They weren't going to build this tower for free. However, since we had only a vague idea of what we would be constructing each week, we couldn't accurately pay them in advance. Nor could we definitively determine what either Con Ed or Silverstein Properties would owe for the work that would be performed.

So I came up with an idea. It was based on a principle that's often rare in business: trust. I got Dan Tishman, the head of Tishman Construction, to buy into it; after all, I had worked with John, his father, for decades before he had retired. I also knew Dan from way back, and he knew me. The Tishmans and the Silversteins had no trouble trusting each other. And Gene McGrath, the Con Ed CEO, to his credit, was willing to buy into my good-faith solution, too.

Here's what we did: Each week representatives from Tishman, Con Ed, and Silverstein would meet in the conference room of my Midtown office. We would go over what had been done during the previous week and we would work out who owed what to Tishman. We would allocate what part of the prior week's Tishman bill was for work on the substation, and what part was the responsibility of the tower. And we continued to resolve the construction bills in this good-faith way, bound only by a gentleman's agreement rather than a written contract, until the substation was completed in September 2003. And, equally amazing, Con Ed and I never had a major disagreement. We never left one of those Friday sessions in my conference room without having agreed on how much each of us owed. This was truly fortunate; if we had been unable to resolve the payment of a single bill from Tishman, work on both the tower and the substation would have immediately come to a halt. And who knows when we would have completed the building? Or, for that matter, if we would have completed it at all.

There was also something else unique about those weekly meetings: I invited the Port to attend. I didn't have to have them in the room. There was no contractual reason that required their observing the negotiating sessions among the three of us. But I wanted to establish transparency among us. I knew I would have to work hand in hand with the Port if I was going to rebuild the Trade Center site, and I wanted them to observe how I operate. I wanted them to know what to expect from me. I wanted them to understand that if we were ever going to accomplish this immensely complicated project of putting up the buildings that would replace the Twin Towers and Tower Seven, we would often need to improvise as we went along. And we would often need to proceed on little more than good faith. I wanted them to trust me, just as I would need to trust them if we were to succeed.

AS THE TOWER STARTED to rise, there were, nevertheless, plenty of people who thought it would never get completed. Their reasoning was grounded in a very practical economic logic. I had only a single signed tenant: Silverstein Properties.

The New York Times very bluntly summed up my shaky predicament. "In the past decade," the paper ominously warned, "no developer has built a major office tower in Manhattan without a tenant for about half the building, typically a requirement of lenders." And the paper twisted the knife further by quoting the president of the city's Economic Development Corporation. "Larry Silverstein," he said with apparent dismay, "is going to build this 1.7-million square-foot tower with no tenants. That's a pretty bold thing to do."

But I was determined to move forward. The insurance money and the Liberty Bonds gave me, I felt, the financial cushion to proceed. Besides, I was, as always, energized by a heady optimism. I believed that businesses would want to return to the Trade Center area if you offered buildings of architectural significance that were also safe and modern. I was putting up a technologically efficient tower, a fully "green" LEED-certified building. And we had brought in an experienced security consultant from Israel to make sure the tower—as well as the seventy-seven-foot-high electrical substation at its base— would be virtually indestructible. I wanted tenants who had qualms about leasing space in a downtown area that had been targeted on two previous occasions by terrorists to understand that they would be safe in this building. No matter the attack, 7 World Trade Center would never come tumbling down.

But there was also another reason for my confidence. It has always been one of my guiding principles never to bet against New York. Sure, I was rebuilding Tower Seven during a time when the commercial real estate market was in a slump and the financial services companies were cutting back. But I've been in the business long enough to have seen that slumps don't last forever. My motivating philosophy has always been "If you build it, they will come—eventually." That is the great thing about New York: people will always want to be here.

Therefore, even while I was struggling to find tenants, I wasn't about to compromise on the rents I wanted. In my judgment, the top floors of Tower Seven were worth $75 to $85 a square foot, the lower floors about $58. When Mayor Bloomberg heard what I was asking, he said the rental prices were exorbitant. He told the press that I shouldn't be asking for a dime over $35 a square foot. That didn't make me too happy at all, especially when it seemed to have an effect on prospective tenants. There were, for example, some serious talks with a major law firm interested in leasing space, only they wanted what struck me as a bargain-basement deal. Maybe they thought they had me over a barrel. And in a way, I guess they did at the time. But I was confident in the product. I was constructing a beautiful building with fantastic unobstructed views. If I waited, I would get tenants who would pay the rents I was asking.

And in the meantime, I'd be protecting the value of a family-owned asset. If I lowered the rent by just $2 a square foot, that would reduce the building's annual income by $3.4 million. It would also lower the value of the total building.

So I waited. And as I had predicted, tenants slowly started to rent space. Moody's, a major financial services company, took a twenty-year lease for fifteen floors. And once they moved in, others followed. Then when it became public knowledge that tenants were paying the sort of rents I had always expected, Mayor Bloomberg had the grace to apologize. "Larry," he conceded during a telephone call with me, "I was wrong."

So with Tower Seven, I did things my way. That's one of the benefits, I have found, to growing old. I had learned to look at things from a different perspective. It might not always have seemed practical in a short-term business sense. But I was being motivated by a lot more than the bottom line. I had begun to understand that if I was going to do something, I had better do it right. After all, there was no telling how many more opportunities I would have.

Then, just as construction was nearing completion, another gamble I had made paid off. The city officially conveyed the title to the strip of land on Vesey Street where my building was already standing. If they hadn't, who knows what would have happened? Fortunately, I never had to find out.

GUIDED BY MY NEWFOUND philosophy, I was determined that this building would be all about the details. That it would be something special. And so I gave a good deal of thought to the finishing touches.

In the rear of the lobby, behind the long reception desk, there was a substantial glass security wall that had been specially designed to absorb a massive explosion. It would recoil rather than shatter. I wanted my tenants to know that they would be safe in case there was another terrorist attack, and I knew that my son and daughter would be working in this building with me, too. Led by Reuben Eytan, whom I had brought in from Israel, where he had a great deal of experience fortifying high-rise structures against bomb attacks, the security engineers had designed something ingenious. But it was also, well, cold and imperious, not the sort of welcome I wanted people to have when they entered the building.

The new 7 World Trade Center rises in Lower Manhattan in 2004.

"Let's do something clever," I decided. "Let's put a piece of art on the wall." But the lobby wall was sixty feet long. What could I find to cover it? After visiting some downtown art galleries, I discovered the work of Jenny Holzer and I loved what I saw. So I commissioned her to do a piece for the lobby. She produced an installation where quotations about the joy of living in New York from the poetry of Walt Whitman, William Carlos Williams, Elizabeth Bishop, and others were projected on the wall in a continuous loop. And it was brilliant. It filled the lobby with color, light, and energy. The last thing you would ever think about is that it conceals an entrance security wall erected to cushion a bomb blast.

Outside the Greenwich Street entrance was a half-acre triangular park on land I'd donated to the city. I could have built on it; after all, every strip of real estate in Manhattan is valuable, and this piece at the time was worth at least $45 million. But I had decided that if I really wanted Tower Seven to be seen as a first-class address, then it needed a dignified entrance. And having a park just outside the lobby would create the hospitable effect I wanted.

There were planters filled with azaleas and boxwood and there was a small grove of sweet gum trees (they look glorious in the fall). And more or less in the center of the park was a round reflecting pool dotted with small gurgling fountains. When I saw the design, I thought it was beautiful. But I also felt it needed something. Like the lobby, it needed an exciting piece of art.

After going to an exhibit on West Broadway, I decided to commission a piece from Jeff Koons. I met with his agent, Jeffrey Deitch, and I made sure they both understood that I needed the piece to be delivered in time for the opening of my building, about a year and a half away.

And so I waited for the preliminary sketches. But nothing came. As the months passed, I thought perhaps he had skipped the preliminary stage and moved on to the fabrication of the sculpture. But when I inquired, all I heard was a litany of "soon, soon, soon."

When the opening had been scheduled and there was still no

sculpture from Koons, I called him, girding myself for another battle. In the real estate business, it seems we're always fighting over one thing or another, but I never thought I would have to lock horns with a world-renowned artist.

And I didn't. Jeff was a real gentleman. He kept apologizing and apologizing, explaining to me that he'd been overwhelmed with work. But I was persistent. I kept repeating how disappointed I was. So finally he said, "Let me make it up to you. Let me loan you something you can put on display. I'll make sure it's delivered and installed in time."

And he did. *Balloon Flower (Red)*, a nine-foot-high bright red stainless-steel sculpture, was sitting in the park's circular fountain for the opening-day ceremony. It looked absolutely, breathtakingly beautiful. (And the sculpture would remain in the park on loan till May 2018. Three years later, I filled the vacant spot with another important sculpture I had purchased, Frank Stella's *Jasper's Split Star.*)

On May 23, 2006, 7 World Trade Center officially opened. It had been quite a battle, but less than five years after its predecessor had been destroyed, the doors of the new building were open. Silverstein Properties was no longer just the sole tenant. Moody's had committed to take 600,000 square feet of office space, and the building was bustling with people. It was the first structure to go up in the new Trade Center complex, and I hoped that this tangible and speedy success would smooth the way as I moved forward. I hoped that it wouldn't be long before I'd start construction on the other towers. On that bright May morning, I was bursting with confidence and optimism. *Things will now certainly pick up speed,* I rejoiced.

TEN

THE EUPHORIA AND CONFIDENCE I felt as Tower Seven opened, however, was fleeting. It didn't take long to discover that erecting a fifty-two-story tower where I was pretty much calling all the shots was one thing. Moving ahead with rebuilding on the remainder of the sixteen-acre site was another decidedly more complex endeavor.

For starters, before I could even begin to contemplate the design and construction of the specific replacement buildings, a basic agreement on the look of the entire new Trade Center complex was necessary (just as David Childs had grasped when he began his preliminary sketches for Tower Seven). This master plan, as it would become known, would need to integrate three elements that were part of the immutable legacy of the old Twin Towers.

For one, the site was a transit hub for Lower Manhattan. Beneath the rubble of Ground Zero was a subterranean world where thirteen active subway lines routinely transported hundreds of thousands of passengers to and from the Wall Street area in the course of a single day. And there was also a PATH line running on an underground loop from New Jersey, continuing on deep beneath the Hudson River, and then into the complex; over 100,000 people used it to travel daily to and from the World Trade Center.

Second, further encumbering any plans for future development of the site, there was a mountain of pre-existing legal agreements and restrictions. I, of course, had a meticulously detailed 1,160 page lease that gave me control of the site where the Twin Towers had stood for the next ninety-four years. There were also hundreds of other tenant leases, multimillion-dollar mortgages, and complex financing documents, as well as easement and operating agreements, all of which had been specifically designed by batteries of careful lawyers and applied to the site. These were all binding legal contracts and they couldn't be cavalierly ignored or dismissed.

Further, and arguably most significantly, this was hallowed ground, the site of an attack on America where nearly three thousand people had died; the rubble held their remains. What was done at Ground Zero would have to honor the victims and their families. It would need to acknowledge the tragedy of 9/11. It would have to be a stirring symbol for future generations of America's defiant response to the terrorists.

The inherited challenges were additionally complicated by the new, post 9/11 reality: the rebuilding would be a de facto partnership—with a myriad of partners. There were, by my count, about twenty public and private stakeholders involved in the process. I was the primary private player; I had the ninety-nine-year lease. On the public side, there was the Port Authority, which owned the land, as well as the governors of New York and New Jersey, and the mayor of New York. At the same time, the federal government, citing emergency powers, had taken control of the site; they needed to be consulted, too.

Then, since this was New York, a boisterous, opinionated town where any proposal involving change was certain to initiate debate, there were local politicians and community groups fervently commenting on each stage of the plans. In addition, of course, there were the families of the victims; what was done on the site where their loved ones had perished was a poignant, heartfelt concern for everyone involved. At the LMDC's first large-scale public hearing in May 2002, a crowd of more than a thousand concerned citizens showed up to have their say. They wanted to make sure, as the formal "Principles for Action" that were drafted at this spirited meeting demanded, that decisions about the site would be made "based on an inclusive and open public process."

And all the while, the always vigilant and outspoken New York press corps—reporters, editorial writers, and architecture critics—hovered over the master plan discussions, often weighing in with a sharp, mean-spirited intensity.

Ada Louise Huxtable, the Pulitzer Prize–winning *New York Times* architecture critic, would, for example, give praise with one sentence and then swiftly take it back in the next. She lauded Daniel Libeskind's initial master plan, but then went on to savage the decision-making that reworked the design: "a dyslexic process (everything backward) that made all the mistakes in the plan book and invented a few." And her ultimate judgment was no less despairing: "What ground zero tells us is that we have lost the faith and the nerve, the knowledge and the leadership" to build a great city "of symbolic beauty and enduring public amenity." Still, the plan had its supporters, too. Steve Cuozzo of the *New York Post* supported the more practical and pragmatic scheme that I'd been advocating. He praised the prospect of the site being "developed by a New Yorker and tailored to the needs of real tenants," a plan that "reflects the American ideal of private ownership sensitive to the public good."

The discussions about the rebuilding proceeded with the combative, energetic spirit of the city itself. Everyone had something to say, a crucial opinion to share. And they all were, it often seemed to me, determined to drown out the voices of everyone else. Yet the battle lines seemed ultimately to coalesce around a single fundamental issue: Should the rebuilding be led by the government or by a private developer?

Yet after two heated years of debate, in mid-2003, Governor Pataki and I, as well as most of the other major stakeholders, had managed to agree on two broad principles. First, it would be disrespectful to the victims and their survivors to build on the footprint of the old Twin Towers. The site of their deaths, Ground Zero for an attack on America, was sacred ground. It should become a memorial. And second, as I had already decided when I had laid out the new Tower Seven, the sterile "superblock" design of the old World Trade Center complex should be eliminated. It would be replaced with a street grid based on what had existed before the two towers had been built: a network of vibrant, light-filled city streets.

Yet while these were both concepts that I applauded, they were also, I quickly made clear to the Port, the governors, and anyone else who'd listen, little more than wishful thinking. The legal reality, as I and my lawyers viewed things, rendered these plans impossible. I had what we felt was a binding ninety-nine-year lease on the site, and it required that I rebuild the Twin Towers exactly where they'd stood—which was exactly where the memorial would be.

The Port and Governor Pataki had a simple solution for this problem: I should step aside and relinquish all rights to the site. All the pre-existing legal arrangements, they suggested, should be cavalierly tossed aside, and the complex should be developed entirely by government agencies. In return, I was offered the prospect of a big check; sums in the hundreds of millions of dollars were dangled in front of me. It was more than enough money to cover what I had already invested. Why do you need all the headaches? they argued. You're seventy-two years old. Why don't you just take the money and sail off into the sunset on your boat, a sizable profit in your pocket?

I never considered the offers for a minute. And while I was convinced I had the law on my side, my commitment to the rebuilding was driven by what I felt was a moral imperative. No one seemed to understand that after 9/11 my involvement with the Trade Center was not primarily about money or profit. There was something larger, more emotional, more significant motivating me. I was determined to show the people who had attacked America that we don't give up. That we rebuild. I was passionately committed to ensuring that what had been destroyed would be replaced with a new, even more vibrant Lower Manhattan. And I also had seen enough about how the Port and the other government agencies operated to know that they would never get the job done. The site would remain empty for decades, while America's enemies gloated. I could not let that happen.

Was I exaggerating the incompetence of the Port and the other government agencies? Was it vanity or ego that had caused me to believe that I could get things done while they would flounder?

Well, look at their track record. Take for one grim example the Midtown Port Authority Bus Terminal, the city's main entry point, a facility that bears the organization's name. It was completed in 1950 for $24 million, yet as early as the 1960s, it had become a center for crime, filled with loiterers, vagrants, and homeless people. The Port renovated and upgraded the terminal over the years, but in the end the results would be accurately summed up by an enraged commuter: "If I die and go to hell, I think it might resemble this."

Then there was the August 1983 collapse of the 13,000-square-foot ceiling at the Journal Square PATH station in Jersey City that killed two people and injured an additional eight. The Port's own official investigation blamed the disaster on inadequate supervision during construction and poor maintenance, inspection, and repair procedures. The Port's chagrined executive director stated it "was a tragedy which need not and should not have occurred. . . . Failure was designed into this ceiling," he conceded.

Or take the city's handling of the rebuilding of the ice rink in Central Park. This was a small project in the scheme of things. Yet the city spent $12 million over six years trying to get the job done— and still failed. Donald Trump, back in 1986 when he was a private developer, had to step in, and in just six months and for $750,000 below his own projected $3 million budget he had the rink up and running.

And what was my track record? Well, in the aftermath of 9/11, when left on my own and not constrained by the Port, I had succeeded in getting the new 7 World Trade Center not only built, but also filled with tenants. And even before the tower was completed, I had the Con Ed substation housed in the building's base operational at a time when the city desperately needed it. Not least, I had completed construction of a beautiful, sustainable, secure world-class building at a price that made sense. A price that ensured a developer, even in a difficult economy, would make a long-term profit on his investment. It was the first building constructed on the site after the terrorist attacks and it demonstrated what I could do.

Therefore, I knew I could get the job done. And I also knew I had a duty to do it. I was prepared to use every legal means at my disposal to fight for my control of the site. I had a contract as thick as a dictionary to back me up, a legal reality that was reinforced by the fact that I was still paying $10 million in rent every month. There was no way I was going to walk docilely away.

IT TOOK A WHILE, but Governor Pataki finally began to understand that I was serious. He began to realize that if things were going to move forward, there would need to be a viable public-private partnership for the rebuilding. And it was the press's prodding, I will always suspect, that brought him to this realization. There had been a flurry of articles blaming him for the fact that two years after the attack, definitive plans had still not been made for the future of the complex. This was definitely not the sort of media attention a man reportedly thinking about running for president wanted. In October 2003, he invited both the Port and me to a summit meeting.

The governor made it clear that he was exasperated. He would not accept, he declared pointedly, continued uncertainty or prolonged negotiations. He had had enough of nonbinding solutions or handshake agreements. There had to be a permanent solution between the Port and me resolving how to proceed, and it had to be formulated without delay. The public-private partnership necessary for the rebuilding had to be explicitly stated.

Impossible, the Port thundered. It would take years to iron out all the issues. And since it was growing clear that the governor—as well as the public—was frustrated with the years of inconclusive debate, there was only one solution: Silverstein should walk away and leave the agency in sole control.

I listened and I didn't get mad. I didn't lose my temper. Instead, I saw an opportunity. The governor was giving us an ultimatum to come up with a plan as quickly as possible. Therefore, I'd do just that—and in the process put an end to the talk that I should relinquish my leasehold. "We'll get back to you tomorrow morning with a plan," I promised the governor. But even as I made that pledge, my confidence was all bravado. *How are we going to get this done?* I wondered.

Nevertheless, I hurried back to the office with my team and we worked through most of the night. We decided to start by specifying what we wanted the plan to convey.

It must, we realized, include all these elements: We had to present an agreement stating that a memorial would be built where the Twin Towers had stood; that the superblock would be replaced by a street grid; and that I, as specified in my lease, still had the right to rebuild 10 million square feet of commercial space elsewhere on the sixteen acres. And these principles had to be articulated in a manner that could not be diluted by lawyerly caveats or further debate. It had to be a short, succinct, and easily comprehensible statement. And it had to be totally binding.

The next morning we presented a two-page letter to the governor. It proposed "an immediate swap." I would voluntarily relinquish my leasehold on the Twin Towers so that their footprints could become the setting for a memorial. In return, I would get a leasehold on five separate sites positioned around the memorial and a reopened street grid. The swap also made it clear that I would have the right to build on the five sites the identical amount of office space that had existed in the two towers before 9/11: "10 million square feet of commercially viable Class A above-grade rentable office space . . . [in five towers] with at-grade lobbies."

On the back of the second page of this succinct letter, we also attached a color-coded diagram superimposed on a map of the sixteen-acre site. It was a vivid, easy-to-understand representation of the swap I was proposing: the footprint transferred to the memorial was in green; what would be my office buildings were in blue; and the location of the new train station and infrastructure was in yellow.

And we made it clear that once the letter was signed, it would be binding. I wanted the Port to understand that this carefully worded swap agreement was not simply aspirational, a plan that might or might not be used as we moved forward. Sure, there would be details that would need to be filled in later, and these would no doubt involve a good deal of spirited negotiation. However, once they accepted these basic terms, there would be no turning back. We would have the gist of the master plan—a partnership between the public and private sectors—for rebuilding the site.

On December 1, 2003, the Port and I both signed the swap letter my team had drafted. The master plan I had proposed took effect immediately. And my right to put up five new buildings—a total of 10 million square feet of office space—became sacrosanct. At last, I felt, the rebuilding could finally begin.

BUT WHAT SHOULD THE buildings look like? Where should they be positioned on the sixteen-acre site? How should their placement be coordinated with the new memorial and train station? And how could the entire complex be designed so that its many separate parts were integrated into a vital, well-functioning, and aesthetically pleasing whole?

These were the questions that the Lower Manhattan Development Corporation, even before the master plan "immediate swap" letter had been signed, had been asking. Remember, as I had mentioned, the LMDC had been set up during the last days of the Giuliani administration, a time when all the polls were predicting a Democrat would become the next mayor. Giuliani and Pataki, however, were hell-bent on their fellow Republicans continuing to have a large say in the allocating of the promised $10 billion (later doubled to $20 billion) of federal funds that would be pouring into Lower Manhattan in the aftermath of 9/11. Therefore, they set up the LMDC to help allocate the bounty of federal reconstruction funds, as well as to advise on the planning. To their surprise, Michael Bloomberg, running as a Republican (albeit without a strong affiliation to the party), won. So now they didn't need the LMDC. But they couldn't just abolish it; that would reveal their political ploy to be, well, a political ploy. They decided they might as well put the organization to some use.

In July 2002, the LDMC announced the first round of what it called the "Preliminary Design Contest." The group would justify its existence by having established planners and architects submit proposals for the future layout of the site. The design mock-ups would be put on display at Federal Hall for the public to view and then comment on their favorites.

And this exercise pretty much summed up the LDMC's role—and ultimate power—in the process: it could espouse lofty goals, set a very public agenda that would get the politicians, the community groups, and the press talking. It had a bully pulpit and could influence the conversation about what should be done. And don't get me wrong; that wasn't a bad thing. It could, in fact, be beneficial. But the group's job of mobilizing people, as Mayor Bloomberg said with a politician's careful tact, "to move forward with confidence and skill when thinking about how to rebuild Lower Manhattan" was certainly not the same as having the final say.

And where did I fit in when these design discussions were being held? I was not consulted, and that left me bristling. I was spending $120 million a year for the vacant land. I had all the construction responsibilities, all the expenses, all the mounting debts. But I wasn't even given the courtesy of being brought into the conversation. It seemed incredible to me that the LMDC didn't reach out to me, didn't try to involve me in the design selection. But I figured it was not the time to make waves; I was still deeply involved in the acrimonious debate over the master plan. I would just stay out of the LMDC's way and see what they came up with.

It was a good thing I did. The entire exercise wound up being an unmitigated disaster. When the mock-ups were displayed, they looked like nothing more than dark globs of clay. You couldn't tell that the shapes had been meant to represent buildings. There wasn't even any fenestration depicted. People who came to the convention center had no context for what they were viewing. They couldn't make heads or tails out of the amorphous clay shapes or appreciate their placement on the sixteen acres. The LMDC had spent months setting up this "design contest," but had never bothered to do the necessary groundwork to help visitors envision or understand how the models might relate to the actual rebuilding. It was an absolute waste of time, effort, and public money.

It was in the wake of this fiasco that Roland Betts took on an important role at the LMDC; there were sixteen board members, half chosen by the governor of New York and half by New York's mayor. Betts had been a college roommate of President George W. Bush, and after Yale he had forged a very successful career in business; he had bought the Texas Rangers baseball team with Bush and had hit another home run with his inventive development of the Chelsea Piers complex along the Hudson River. And now Betts was made chairman of the site planning committee for the LMDC. His specific assignment: to clean up the mess the LMDC had created with their earlier public presentation.

In August 2002, Betts launched a new competition, the Innovative Design Contest. This time he made sure the designers and planners understood their mission. They were given instructions to build precise architectural models, to show trees and streets, to give a sense

Daniel Libeskind's architectural model for Memory
Foundations, his proposal for the new World Trade Center

Proposal by THINK Design, a group including Rafael Viñoly,
Frederic Schwartz, Ken Smith, and Shigeru Ban

of the vibrant neighborhood that would be created on the Lower Manhattan site. There were over two thousand entries. And from this group, six semifinalists were selected and their visions displayed in the Winter Garden of the World Financial Center to a now engaged public.

At this point in the process, I figured I had better get involved in the discussions that would select the final plan; after all, my lease specifically stated that I had the right to select an architect should the Twin Towers need to be replaced. I told my lawyers it was time to assert Silverstein Properties' role. They sent a letter to Pataki, which he ignored—until it appeared in the papers. At that point, the governor undoubtedly understood that not only would I continue to assert my contractual rights for the planning of the site, but that I would also find a way to make sure the press was aware of what was on my mind. Pataki now conceded that I could have a seat at the judging table, too.

In addition to me, the decision-makers included the LMDC board, the mayor, the Port, and—the first among supposed equals—Governor Pataki. We looked at the finalists' plans and were instructed to judge them based on twelve criteria the LMDC had established. These included price, public response, vision, connectivity, public space, and how the victims of the attack would be memorialized. These were all worthy, high-minded standards, but also incomplete. As someone who had actually put up commercial office towers, I knew you had to pay a great deal of attention to feasibility. Could the high-rises be constructed as designed on a realistic budget, at a cost that would allow you to attract corporate tenants at a competitive rent? What good would the prettiest picture be if it couldn't be translated into a tower that would actually get built, that would one day be teeming with activity?

Those were the troubling concerns I had when I studied the design presented by one of the finalists, Daniel Libeskind, a Polish-born American citizen now living and working as an architect in Berlin. Libeskind was a very engaging guy, a wonderful storyteller.

The Freedom Tower by Daniel Libeskind, inspired by the Statue of Liberty

And he looked the part, too, with his shock of silvery hair, his all-black outfits, and his thick-framed glasses. He had designed the Jewish Museum in Berlin, for which he had received a good deal of well-deserved praise; its jagged, brutalist edges were an effective symbolic reference to the raw horror of the Holocaust extermination camps. But that was the only building of his that had ever been built. And it was just four stories. I wasn't convinced he had the sort of practical experience or expertise that would be needed to direct one of the largest commercial construction projects in history.

When he presented his master plan for the Trade Center, it had lots of doodads, a conglomeration of angular buildings of descending heights that would straddle the footprints of the original towers. And sure, it was impressive, and smart in a conceptual way. But when I studied his design with my builder's eye, what I saw was his vision, not a master plan. It wasn't something that could ever get built.

I wasn't the only dissenter. Roland Betts, a developer himself and the LMDC's chief voice in the discussions, favored Rafael Viñoly's design. And it was indeed beautiful, a shimmering latticework of two interconnected towers. But it, too, in my judgment, was simply visionary, not a practical guide for the construction of 10 million square feet of leasable office space.

But it didn't matter what either Betts or I thought. Sure, we had votes, but they didn't count for much. The governor had the final say, and he seemed determined to pick Libeskind. Pataki was particularly taken with the 1,776-foot height—of course, a deliberate patriotic symbol—that Libeskind had questionably anchored in a far corner of the roof of the highest tower. In fact, the entire angular seventy-story tower, an intricate assemblage of sharp lines and precise edges, had been plunked down on the site without paying any attention to the existing slurry wall, the train tracks and tunnels below grade, or the depth of the bedrock. Or, for that matter, the security risk of putting it perilously close to the adjacent West Side Highway. All that apparently mattered to Libeskind and the governor was a modernistic design that would look good when reprinted in *The New York Times.* The governor had christened this building "Freedom Tower," and that, too, gave me pause. How many tenants would want offices in a skyscraper that was tacitly advertising itself as a target to a new generation of terrorists?

But what could I do? I figured I would bide my time. I had already been working with David Childs on a design for a tower with 3 million square feet that would be the commercial centerpiece of the site; I would continue to proceed quietly with that collaboration. Libeskind's quixotic design, I predicted, would self-destruct as the process moved further along.

Still, I was encouraged by the meeting I had with Libeskind. Like many of the finalists who realized that ultimately they would have to work with me, he came by my office to introduce himself. And he brought his wife, Nina, with him. I liked that.

I also liked what he had to say. "If I'm chosen, one thing I can promise you," he pledged with earnest conviction, "is that I'll give you full cooperation. We'll work together very closely. I'll respect your needs because you're the developer. I'm not going to be a problem for you. I'm going to be an asset."

"Great," I said. And I just hoped he meant what he said.

WHEN LIBESKIND WAS CHOSEN as the competition's winner, there was a great deal of fanfare. He was publicly anointed by the governor and the press as the visionary architect who would design the World Trade complex. Now I began to worry; much more credit was being given to this man than made any sense. He wouldn't be able to do this. It was too large an undertaking.

It was in the midst of his being lionized by an adoring public that Libeskind gave a talk to the Association for a Better New York, a well-respected civic group. There were three hundred people at this breakfast in the Plaza Hotel, and a report quickly came back to me.

"Exactly how I laid out these buildings is exactly how they're going to be placed," he told the audience, according to the story that was related to me. "And exactly how I designed them will be exactly how they'll be built." The adoring crowd erupted into cheers, applauding his commitment and ambition. But when I was informed, I began to think that the far-fetched Libeskind show had gone on for too long. And since I'd be paying the bills for the buildings, I had better put an end to it.

So I invited Libeskind to my office. And like his initial visit, he brought Nina, too.

"First, congratulations," I began with tact. "I think it's terrific you were chosen. A real honor. And I like the overall design you did. In fact, it's quite similar to the layout we had done with David Childs. Enormously similar, in fact."

Then I showed him Childs's design. It, too, had five towers scattered around the complex, with the centerpiece building positioned northwest of the old Twin Towers. It had also envisioned a memorial

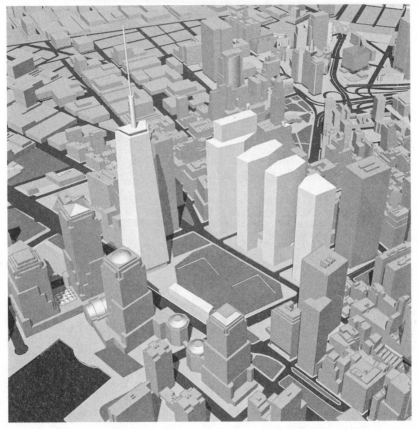

A 2002 massing study by Beyer Blinder Belle commissioned by the
Lower Manhattan Development Corporation (LMDC)

built on the site of the two collapsed towers and had planned for a
transit hub.

"There just aren't too many ways you can lay this site out," I
continued. But at the same time I kept the thought to myself that
there had not been any need to waste valuable time and god knows
how much federal money (the corporation was funded by the U.S.
Department of Housing and Urban Development) on the LMDC's
series of futile competitions. What would be the point? Libeskind
wasn't to blame for that.

Instead, continuing along in my understanding and conciliatory way, I explained that I wanted us to work together—to a degree. "Give us your plan. We'll use it as a starting point. But I'm going to have David Childs from Skidmore execute the final design. He's experienced. He's built office towers before. He knows all the ropes. I'm paying the bills, and I want him to be the lead architect."

Libeskind didn't want to listen. "No, no, no," he protested. "I'm going to design all the buildings."

He was clearly adamant, but nevertheless I thought I could still reason with him. "Dan, you've never designed a high-rise building in your life. And now you're going to develop five of them? Including one of the tallest in the world?"

It was not, however, a time for reason.

Nina now spoke up. "Dan's a quick learner," she said. "In one afternoon he can learn everything he has to know about high-rise office buildings."

"What are you talking about?" I asked matter-of-factly. I was determined to keep my cool.

"There were three hundred people in the Plaza the other morning who now expect Dan to design the site, to design its tallest tower, and that's exactly what he's going to do," she announced. "And that's exactly what you're going to build!"

For a moment I couldn't decide what was more astounding, her naïveté or her rudeness.

"You're telling me?" I finally challenged, my affability now gone. "It's my property. It's my money. And you're telling me what I'm going to build?"

Then I took a deep breath and reined in my anger. "Look," I said, hoping that logic might still persuade her. "Suppose you had cancer of the brain and needed a brain surgeon. Would you go to someone who's a 'quick study'? Who would try to learn everything in an afternoon? Or would you go to a specialist?"

"It's not the same thing," she tried.

"Let me tell you something," I said firmly, all pretense of conciliation now abandoned. "It is the same thing. And as far as I'm concerned, you're threatening me. This meeting is over. I want nothing further to do with you. You can leave."

"Don't get so excited," she tried. But her patronizing tone only added fuel to my burning temper.

"This meeting is over. You can leave." But not before I added with a steely promise, "I'm going to design these buildings. I'm going to build them. Not you."

They walked out seething. But I can assure you their rage was nothing compared to mine. I was committed to the rebuilding, and I knew the process would only succeed if there was a viable architectural design, a configuration that made both engineering and commercial sense. I was not about to let the entire future of the site be destroyed by a quixotic plan that could not be built. I recognized Libeskind's talent and his artistic vision. But this was not the appropriate project for his flights of fancy, however beautiful they looked on the page. And I could not work with someone who seemed uninterested in working out a reasonable compromise. There was too much at stake.

I DIDN'T HAVE TO wait long until Pataki called. I had expected to hear from him, but I had not anticipated what he'd say.

"Listen," he began, "let Dan and David Childs design the buildings together."

"Together?" I echoed. "That'd be impossible. These guys live in different worlds. David spent his life designing high-rise office buildings. He's recognized throughout the world for his many successful projects. He's created magnificent structures that actually get built. Embassies. Corporate headquarters. Banks. Office towers. David has been conceiving big projects for thirty years. While Libeskind hasn't built anything over four stories. And it's his entire construction portfolio. That single building."

"Larry," the governor said plaintively, "help me."

So, when my temper cooled, I thought about it. Heading into the many battles that undoubtedly lay ahead, it wouldn't hurt to have the governor owe me one. It would be an IOU of sorts that maybe I would be able to cash down the road. And I knew that David's skill and expertise would triumph over Libeskind's bravado in the final building design. The challenge, though, would be to convince David to work with such an idealistic and too often impractical collaborator.

"Can you please get together with Dan," I tried when I spoke with David. "The building will be yours, but you'll give him some credit."

And David, an old-school gentleman, after some more prodding acquiesced. He agreed that the Trade Center needed to be rebuilt, and if his cooperating with Libeskind was the price he would have to pay, it would, in the scheme of things, be a small sacrifice. "I'll try

The hybrid design of the Freedom Tower by SOM and Studio Libeskind

to work with him," he agreed without much enthusiasm. "How bad could it be?"

It turned out to be six months of hell. At the time it seemed as if Libeskind fought with Childs tooth and nail over every aspect of the design. It was impossibly difficult, but they finally got the tower designed.

I looked at the plan, and it was god-awful, just terrible. And in addition it was an offset tower, which I knew would never work on the site. Not with the slurry wall at its base. Besides, it would create too many security concerns.

"David," I asked, "how in the hell is anyone going to build that? How is it going to stand? And that tall antenna on the roof. In the corner? God forbid anything happens. That there's an attack on the building. It'll just come right down. You can't have that."

"Larry, that's what Libeskind wants," David explained.

So I thought about it and made a strategic decision. "Okay, we need to move forward. We can't spend any more time fighting over the design. Let's just get something out there the Port and the governor can approve. When it comes time to do the actual construction, we'll get some industry people, engineers, to tell the government it's not smart to build it offset. That for security reasons the tower will need to be supported by something in the middle, a solid structural core. That the antenna will need to be moved to the center of the roof. Then we'll redesign the building."

So I kept quiet, just biding my time as the Port and the governor approved the design. In fact, a jubilant Pataki declared, "This is not just a building. This is a symbol of New York. This is a symbol of America. This is a symbol of freedom."

Me, I had put up enough buildings to know it was a symbol of naïveté. It was a design that would never get built. But I decided that it wouldn't make sense to say what was on my mind. As soon as I did, I would once again become the focus of the story. I did not want to give the politicians and the press another excuse to villainize me. Not that they seemed to need one.

. . .

INDEED, I HAD BEEN completely unprepared for all the very personal attacks that had been hurled at me. Sure, I knew from the start that the rebuilding would be a long, complicated process, but I had never anticipated that my motives would be impugned, or that the criticism would be so vituperative. In my innocence, I thought just about everyone would see the wisdom—and the national honor—in rebuilding the Trade Center site into a thriving commercial hub. I did not have a single doubt that this was what America must do. But since a lot of money was involved—commercial office towers are most fundamentally a business proposition—it seemed people had their knives out for me.

It wasn't just the anti-Semitic lunatic fringe. I could shrug off the rantings of the wild-eyed conspiracy theorists who claimed I had been involved in some kind of crazy plot to destroy the Trade Center in order to obtain a windfall of insurance money. This kind of venom had its origins deep in the toxic myth of the greedy Jew, so the hell with them! I wouldn't deign to argue with them.

But it was much more difficult to ignore the attacks made by the family members of those who had perished on the site. I understood their pain, but I did not understand their hostility toward me. I was not attempting to desecrate the hallowed ground where their loved ones had died. Rather, I was hoping to create a somber memorial and a business complex that would be an active, vibrant memorial, too. When many of these families spoke out at community meetings or to the press claiming that all I cared about was making money, well, that hurt. They had completely misunderstood my motivations. If I had wanted simply to make money, I could have reaped a sizable profit and saved myself years of hard work and aggravation simply by taking the government's lucrative offers to walk away.

In my desperation, I met with representatives of the survivors. I tried to convince them of how important rebuilding the site would be for the entire nation. But many of them could not be moved. It was a terrible situation, and my heart went out to them, but I also felt I had to do what was necessary for the future of the city. There was room on the site for a memorial and for the towers that would help revitalize Lower Manhattan. I wanted to do right by the families and for New York and its future.

Meanwhile, it also seemed like the politicians had ganged up against me, too. Bloomberg had said "it would be in the city's interest to get Silverstein out." (Later he would apologize.) Dan Doctoroff, the deputy mayor in charge of economic development, then doubled down on his boss's initial antagonism, telling a city council hearing that I would ultimately default, putting up at most a single building after pocketing a fortune from the insurance proceeds. My forging ahead would be "a disaster for the city, state, and Port Authority and a blemish on the memory of those who lost their lives on 9/11." And that cruel attack was repeated by Pataki. "Larry has betrayed the public's trust and that of all New Yorkers. We cannot and will not allow profit margins and financial interests to be put ahead of public interest in expediting the rebuilding of the site of the greatest tragedy on American soil." The Port's chairman piled on, too, although he carefully kept his knife in the sheath. "I'm happy to work with Silverstein Properties," he said. "And I'm happy to work without Silverstein Properties."

And although the politicians' attacks were wounding (and often unexpectedly vitriolic), I was determined not to fight back. I tried to keep my eyes fixed on the job that needed to be done. I knew that if the site was going to be rebuilt, we would all have to sit down together and work things through. What would be gained by my responding to fire with more fire? In the end, the only thing that mattered was my getting the towers built. And so I was determined that despite all the name-calling, I would take the high road. I wanted to make sure that I could come back the next day and work with the politicians. I kept my peace, even though I was often fuming.

Much of the press, too, had decided I would be their scapegoat for all the delays and the confusion as the development process dragged on and on. The *Daily News*'s front page shouted, "Larry's got to go." *The New York Times* ran an editorial headed "Greed vs. Good at Ground Zero." Guess who was cast in the role of Greed?

After reading that, I thought that maybe I just needed to explain things better to the paper of record. I mean, I'd spent my life reading the *Times* each morning. So I managed to get the publisher, Arthur Sulzberger, Jr., to invite me to a Publisher's Lunch, as those coveted meetings were formally called, at the newspaper. It'd give me a chance to talk to editorial writers and a few selected reporters.

As soon as we sat down, I could tell the mood in the room was stony. Their questions were very pointed, very accusatory, but I answered them all. I took great care to explain things carefully, to give them a sense of what was really going on behind the scenes. And by the time we were sipping our coffee, I felt things had changed. They were laughing along with me, smiling at my observations. I felt that I had won them over, and that in the future they would write, if not with sympathy for my predicament, at least with more objectivity.

Later that week, another editorial ran railing against the delays in the reconstruction of the Trade Center. And once again I was put straight in the crosshairs as the principal culprit. They pointed to my lengthy legal fight to recover the insurance proceeds from the attacks on the Twin Towers—but never mentioned that I had been the one paying the exorbitant insurance premiums that entitled me to the money I sought. I couldn't understand their criticisms. That's why you buy insurance, isn't it? To protect yourself in case of disaster. It was the same sort of prudent action every homeowner takes. You pay your insurance premiums each month because you expect that if your home were to go up in flames, the insurer will give you the money you need to rebuild. The only difference in this case was that the "home" had initially cost $3.2 billion.

And they also railed against "[Silverstein's] penchant for endless negotiations [that] often left him at odds with city, state and Port Authority officials." *What did they want me to do?* I wondered. *Go along with a lot of bad, half-baked ideas for the rebuilding?* I wasn't arguing for argument's sake. I was doing what I thought was necessary to make sure the job would get done. And a corollary to that more fundamental motivation: I was also the leaseholder who would have to live with what was decided. I would have to find the tenants and get the towers up and running. I needed the complex to be a commercial success and so, sure, I was zealous in my battles.

In fact, the officials now wanted me to invest more of my own money in the project. "He's playing with the house's money and, nothing wrong with that, but let's not feel too bad here," said Mayor Bloomberg.

Talk about chutzpah, I thought sourly. They wanted me to put more funds into the project and at the same time have less say. That didn't strike me as the sort of deal any businessman, including Mike Bloomberg, would make.

But I said to myself, *Forget about it. Forget about the press attacks. Forget about winning people over.* All I could do was get the job done. And if I succeeded, if the new towers rose from what had once been rubble and reached up to the sky, if the complex was alive with office workers and visitors to the memorial, if downtown became a bustling neighborhood, then perhaps my critics would at last begin to comprehend what had been driving me.

ELEVEN

WHILE ALL THIS WAS going on, I was still fighting with the insurance companies to get the bulk of the money I was owed. It had been a long-running battle that had started years earlier, almost immediately after the destruction of the towers.

When I had purchased the leasehold, the Port had insisted that I increase the amount of insurance from the meager $1.5 billion they had on their entire portfolio of properties to a sum equal to what I'd be paying—about $3.2 billion. It was the largest amount of coverage ever purchased for a single office complex and, in the end, I had to buy policies from twenty-five different companies, an elaborate collection of domestic and foreign insurers and reinsurers. And I had been waiting for years for the money I was owed.

In truth, I now needed it desperately. And it wasn't that I just wanted the insurance proceeds to pay the $120 million annual rent on the complex. I had come to the unnerving conclusion that everything I hoped to do depended on it: without that money, I could never afford to rebuild.

But I soon discovered that getting paid would not be easy. Rather, filing the claims initiated a long, expensive, and infuriating legal struggle. It would drag on bitterly for years, and in the end I would learn a very disheartening lesson: When you pay massive premiums for enormous coverage, all you have bought is the right to sue the insurance companies. They were never going to pay the full amount due unless I took them to court.

THE FIRST BATTLE IN this long and acrimonious war broke out when, on my instructions, Wachtell, Lipton had summoned representatives from all twenty-five of the insurance companies to a meeting in the law firm's conference room. It was not long after 9/11, and the insurance companies had largely responded to my initial request for payment by chanting an identical mantra: "You'll never build. You'll never build. You'll never build." I wanted to set them straight.

Herb Wachtell introduced me by saying, "Larry, tell them what you're going to do."

So I told them. And I didn't talk in vague, wishful terms. I shared a concrete, very specific plan. I made it clear that I was determined to rebuild the complex and presented my vision of five towers that would be arranged around a 9/11 memorial. This would be, I said, my unwavering guiding strategy for the future.

They seemed shocked. They refused to believe I was serious. And their reaction very swiftly turned from incredulity to downright anger. It wasn't long before they were fuming, bristling with undisguised hostility.

It's going to take years to get all this done, they challenged. Why would you even want to think about doing this?

My initial response was to point out the patent absurdity of their question. "You can't leave the site a ruin. You can't leave it fallow. A big hole in the middle of downtown Manhattan? This has got to be rebuilt."

And while today this might seem like an absurdity—a vacant muddy pit in the heart of Lower Manhattan—you need to remember that this was exactly what it would be for more than five full years, until construction would start in earnest. This was what had existed before the complex became the linchpin for a thriving community bustling with stores, restaurants, hotels, and schools. And without the insurance proceeds, how much longer would the process have taken? There's no telling. The site might very well have remained a scar on the cityscape for years—decades?—longer. Or Rudy Giuliani's vision of a sixteen-acre memorial filling the entire plot would have, out of inertia, triumphed, and the complex would never have been brought back to life.

But they were unmoved by my heartfelt warnings. Instead, they kept hurling their various arguments in an increasingly persistent barrage. Number one: You are in your seventies. Number two: Why spend the rest of your life bogged down in a major legal fight with us? And number three: There must be some amount of money that will make you walk away. They kept dangling this plump carrot: Name a sum. We will give you any reasonable amount you want to settle. Take the money and run. Forget about the World Trade Center.

It was what many people in government, both city and state officials, had been suggesting to me, too. Hand the property back to them, accept a fee, and they would find a group of developers to come in and take it over. It was too large an undertaking, they said, for a single company, especially when it was a privately owned midsized development concern. And at the same time many of my "good friends" in the real estate business were beating a path to the governor's door demanding, Why should Silverstein have this whole thing? We are more qualified, more likely to get the job done. Some of these "good friends" were also calling me directly, offering their assistance with, they assured me, my best interests at heart. "Are you sure you want to do this alone? You sure you don't want some help?" they coaxed.

By now I had heard enough from everyone. I decided to make my position clear once and for all. I was addressing the room full of insurers, but I might as well have been speaking to the hectoring government officials and my meddling industry friends.

"It's not a question of money," I began coolly, but my temper grew as I spoke. "There's no amount of money that would persuade me to walk away. Here's what you must understand—I'm a New Yorker. I spent my whole life in this city. Not to rebuild this complex? Are you kidding me? How could I walk away from this? What's the matter with you guys? Our country's been attacked and you want to leave it like this? Are you nuts? How could I live with myself? I am going to rebuild the Trade Center and nothing is going to stop me!"

As they grappled with the implications of that firm pledge, I proceeded to let my second shoe drop. And it landed with an even bigger bang.

ON THE CALM LATE summer weekend before the storm of 9/11, Klara and I had, as I previously recounted, docked our boat in Sag Harbor and then gone to dinner at Herb Wachtell's East Hampton home. And in the course of that pleasant evening, Herb had mentioned that he had just returned from a court in Albany, New York, where he had been arguing an interesting case. The issue being litigated, he explained, involved determining the number of occurrences covered under a single general liability insurance policy. This was, he said, a hot topic in insurance law. He cited, for example, a case that had made its way to the Florida Supreme Court, where the number of shots fired in an altercation in a restaurant was successfully used to fix the insurer's exposure: each individual bullet was ruled a separate event and therefore multiple payments were due. It all had seemed at the time a bit arcane for weekend dinner party conversation. But within weeks, everything I had heard that night became of great importance to me.

Because in the still raw days after 9/11, Herb came to me with a dramatic legal insight that had been largely inspired by the case he had argued in the Albany courtroom. And it offered, I at once realized, the prospect of a realistic path toward my rebuilding 10 million square feet of commercial space.

Herb's contention was this: The attack on the Twin Towers was two discrete events. Two separate planes had hit two separate buildings at two separate times. Further bolstering his argument, each tower had collapsed independently of the other, crashing down at different times, too.

And since there were two distinct events, I was entitled to twice the face value of the policies. Instead of recovering $3.2 billion, I should receive $6.4 billion. It would be the largest insurance claim in history. Yet I knew I would need every cent if I was going to rebuild the Trade Center. I was, in effect, the insured party who wanted to rebuild his "home," and this was what it would take.

When I shared this with the representatives of the insurers, they hit the roof. I had expected that and told them that I was adamant. I would sue, if necessary. As it happened, one of the major insurers, Swiss Re, sued first.

There would, in fact, be a long, tedious series of trials and appeals, and each was an expensive ordeal. The legal issues ultimately boiled down to the specific forms that had been used when the original insurance agreements had been written by my broker. The insurance companies argued that they had agreed to a boilerplate contract, what was called the WilProp form. This defined "occurrence" to mean "losses or damages that are attributable directly to one cause or to one series of similar causes."

My contention, as argued by Herb, was that the insurers had agreed to use a form issued by Travelers Property Casualty Corporation. This document did not define "occurrence." And since it had not been defined, precedential case law should be applied that had established multiple incidents as the legal standard for reimbursement.

Yet there was another complication to this already nuanced legal issue. On September 11—remember, this was just six weeks after I'd acquired the leasehold—final insurance contracts had not been signed. All that had been worked out was a series of preliminary agreements, or, as they are generally called, binders.

Nevertheless, it had been clear to me—and this was the thrust of Herb's argument—that by September 11 the negotiations had sufficiently progressed so that the insurers were poised to sign the Travelers form. However, as this document didn't have a definition of occurrence, the court had no choice but to enforce the standard implicit under New York law. And since two separate incidents involving two separate planes had clearly brought down the towers, I was entitled to collect for both. The insurers owed me almost $7 billion.

Only the initial trial verdict didn't see it that way. The proceedings were extremely contentious, extremely confrontational. It seemed to me that we were constantly battling against a judge who believed I had simply come up with a clever idea to get money I wasn't entitled to. It didn't seem to matter to him that I had what I felt was a binding contract—morally as much as legally—that required I rebuild the towers. It didn't seem to matter how fervently I pledged my commitment to the reconstruction. He apparently didn't believe my words were sincere, and he appeared to do his best to exclude any mention of my plans for the future of the site. No matter how my attorneys tried to get testimony admitted that would put the issues in the relevant context of revitalizing Lower Manhattan, the judge moved quickly to strike it down. His narrow legalistic approach, I felt, ignored the larger issue of the future rebirth of the city. And when the verdict came down against me, I wasn't surprised. But I was deeply disheartened, for myself and for New York.

There was another trial, and then an appeal. In these proceedings, too, crucial evidence was capriciously, in my opinion, excluded by the judges. It was even charged that I had violated a court order not to discuss the case in the press because I had given an interview about my plans for reconstruction. The judge threatened to hold me in contempt. It was as if, I began to feel, the court had it in for me personally, that they were putting me, my very character, on trial. Sure, we were arguing a novel, perhaps even controversial legal claim, but I grew convinced that for certain judges their antagonism was rooted in seeing me as the personification of a simplistic cliché—the rapacious New York real estate developer.

I realize as I write these words that a cynical reader might begin to think that I was inventing demons, perhaps even grossly misrepresenting the fundamental animosity of the courts. Yet as I made my way through courtroom after courtroom, this sort of prejudicial behavior was precisely what my lawyers and I had to face. And if it sounds incredible, well, that's because it was.

But by December 2004, after spending millions in legal fees, I had finally won two major courtroom victories. Two federal juries upheld my original contention that the attacks on September 11 had been two separate incidents for insurance purposes. The verdicts were welcome, but not all I had hoped for.

The decisions effectively divided my insurers into two groups. The underwriters who had used the boilerplate contract—the WilProp form—provided by my broker would be responsible for only a single occurrence. However, the nine insurers who had not used this form were liable for two separate occurrences. They would be obligated to pay more than double the face value of their policies—$2.2 billion. A total payout of $4.68 billion was due.

It was a sum that wouldn't nearly cover the estimated $7 billion cost of the new towers. But it would effectively persuade the Port and the other public agencies involved that I, now backed by $4.68 billion in insurance proceeds, was essential to the process. For the reconstruction to proceed, they would need me as much as I would need them.

And so I eagerly awaited the check.

AND I WAITED, AND waited. Eight carriers—a lineup that included Allianz SE, Swiss Re, and the Traveler's Property Casualty Corporation—who were liable for $2.1 billion said, in effect, "not so fast." They claimed that the 2004 trials had only decided the number of occurrences.

The verdicts had not established the replacement value of the Twin Towers. Before they would pay a single dollar, they insisted, it would be necessary to determine what it would cost to build two sparkling new towers *exactly* identical to the buildings that had been destroyed.

In September 2004, three arbitrators were appointed to try to figure out what the bill would be if the World Trade Center towers that had been standing on the morning of September 11, 2001, were to be built today. To make that calculation, the insurance company lawyers were set on examining the cost of every single item used in the construction of what were then the two largest buildings on the planet. They went over the Twin Towers rivet by rivet, bolt by bolt, concrete block by concrete block. Each building was analyzed down to the last screw. The insurers' legal team even subtracted hypothetical depreciation from the original cost of every item that would be needed to replicate each building. And, of course, my legal team had to monitor them throughout this laborious exercise, checking to make sure their appraisals were accurate rather than self-serving.

It was an onerous, argumentative, and expensive process. One of my lawyers called it "a litigation within a litigation" and "an extraordinarily intense, detailed proceeding." And throughout this accounting, the insurance companies continued to contend loftily that until the entire appraisal was completed, they could not calculate what they owed. But I believed their actions were motivated by another, baser purpose: their painstaking strategy had been deliberately designed to take time. Their plan seemed to be to stall, stall, and then stall some more. The companies would do whatever they could to delay paying the billions the courts, at the end of all the trials and appeals, had ruled I was owed.

Frustrated, I went to the New York State superintendent of insurance, hoping that he would be able to provide the arm-twisting that would apparently be required to force the insurance companies to abide by the courts' verdicts. However, he blithely passed the buck. "As soon as the governor gives me the green light, I can collect for you," he offered.

Naturally, I then quickly got in touch with Pataki. Only the governor was similarly evasive. "I need to think about it," he said.

Five months later he was still thinking. So I tried again. "What's wrong? What's preventing you from giving your superintendent the green light?"

"I'm concerned about setting a precedent," he replied.

A precedent? We never had a precedent like 9/11. I was livid, but I kept my thoughts to myself and let him continue.

"I just don't know if I should be doing this," the governor went on.

I listened politely, but all the time I'm shouting to myself, *What's there to know?* In the past, Pataki had repeatedly said to both me and the press that the rebuilding of the Trade Center was crucial to his legacy. Well, now he had an opportunity to help guarantee that it would get done—only he was hesitating. All he had to do was enforce the verdicts of two separate federal juries. What was going on? What was his real reason?

It wasn't until later, after giving our perplexing conversation additional thought, that I finally put all the pieces together. It suddenly seemed clear to me why his previous enthusiasm for getting the site expeditiously restored had been replaced by a pensive procrastination. At last I grasped what I believed was the crucial difference between then and now.

As with so many of his decisions during the rebuilding, Pataki's actions were, I came to believe, influenced by his plan to run for president. After being three times elected the Republican governor of New York, and having been chosen to give one of the principal speeches nominating George W. Bush for his second term as president at the Republican Convention in September 2004, his eyes were set on the heady prize of national office. That was no secret; Pataki's presidential ambitions and plans to enter the Republican primaries for the 2008 election were bandied about in all the papers. And it occurred to me that this was possibly why Pataki was suddenly looking at the insurance companies differently. Perhaps he now saw them as potential contributors to his national campaign. He wanted their support. And rather than antagonize the deep-pocketed insurers, I speculated, he would shrug off his obligations to the city and state of New York. He wouldn't enforce the orders of the federal courts, and in return, he hoped they would write checks for his campaign. If true, it would be horrendous, but there was nothing I could do. Pataki was the governor, and he alone held the power to force the insurers to pay.

THEN IN JANUARY 2007, Pataki was finally gone, and Eliot Spitzer became governor of New York State. He had been in office only days before I called him. (As I said, I am tenacious.)

"I need your help with the insurance companies," I dived in after only a brief preamble. Spitzer had been an aggressive state attorney general before running for governor, and I was counting on him to still be willing to challenge powerful interests. I continued: "I can't collect the money they owe me. They're stalling. And if I can't collect, then we can't rebuild the Trade Center."

"Come on over," he said without hesitation.

When I met the governor, he brought Eric Dinallo, then the acting state insurance superintendent (he would be confirmed a few months later) into the meeting. I explained to them that it was six years after the 9/11 attacks and three years after the federal verdicts awarding me $4.68 billion, and I was still waiting for the money. They promised to look into it.

And they did. In March, Dinallo summoned the representatives of the insurance companies and my people to a meeting in his Lower Manhattan office. When Dinallo had previously worked with Spitzer in the state attorney general's office, people called him "the Hammer." After this meeting, I understood why. He hammered mercilessly away at the insurance company representatives.

He told them, as the events were related to me, that their refusal to make the payments that were due was a "black eye" and "a disgrace" to the entire industry. Even better, he reinforced his criticism with a viable threat: If there wasn't a settlement within four weeks, he would launch an investigation into all their insurance claims-handling practices. If that happened, he warned, they might very well be shut out from writing insurance policies in New York State, the financial capital of the world. It was a consequence that, the companies immediately understood, could wind up costing them a lot more than the billions they owed me.

The insurers were angry as hell; they had been convinced they would never have to pay. Nevertheless, in the weeks that followed the ultimatum given in Dinallo's office, settlements were quickly negotiated with several of the companies. One group of insurers and reinsurers that collectively owed a total of about $1.1 billion, though, asked us to come to Zurich, Switzerland, to work out the final arrangements. So Marc Wolinsky, the Wachtell litigator who was directing my attempts to collect; Mike Levy, my chief financial officer; and Albert Rosenblatt, a retired New York State Court of Appeals judge who had been appointed to mediate the insurance dispute, took the red-eye over to Switzerland.

They arrived in time for the scheduled preliminary lunch with the three representatives of the French-based company. It was in the dining room of the hotel Baur au Lac, a very grand and elegant salon whose windows offer a picture-perfect view of Lake Zurich; a conference room had been reserved for the meeting that would follow. And everything went well, my team having a seemingly cordial discussion with the insurance executives. After about an hour and a half of talk, however, I was told the insurers said they would like to confer among themselves. Please excuse us, they said. We're just going into the lobby. We'll be back shortly.

So my guys waited. And according to the accounts they shared with me, this is what happened: They were sitting in the dining room, looking out forlornly at the lake. An hour passed. Then another. They texted the insurers. They called their cells. No response. Bored, they got up and went for a walk around the carefully manicured hotel grounds. Still no word from the insurers. At around midnight, a bewildered Wolinsky sent the insurers a final email: If we don't hear from you, we're getting on a plane in the morning and returning to New York.

By then the insurance executives were already back in Paris. They had capriciously walked out of the lunch (leaving the bill for my people) and gone straight to the airport. Their story about heading into the lobby for a discussion had been a complete lie. They just left my people waiting for them. And they never said a word. Never apologized. It was incredible. A bad way to do business, and a bad way to treat people.

By May 2007, with my pushing and Spitzer and Dinallo resolutely backing me up, I finally collected. I wound up getting all the money I was entitled to, one hundred cents on the dollar. (Actually, that's not precisely accurate. By the time the negotiations were concluded, I was $12,000 short of the total sum the courts had awarded me. Still, I'd received about $4.68 billion from the insurance companies.) I also tried to collect the interest the insurance companies had earned by holding on to the money for six years rather than honoring their obligations. They had made, it was estimated, about $1 billion on the float. It was a colossal sum, and they had earned it by doing everything they could, using every delaying tactic their lawyers could find, to avoid paying me. It was disgraceful. But, to my bewilderment, I got nowhere. The courts didn't want to hear my argument.

Nevertheless, after all the years of politicians showing up at Ground Zero for fatuous photo ops, after all the heated debate about what should be built on the site, it looked like things could start to move forward. I finally had the funds that would allow me to start work in earnest on the commercial centerpiece of the complex, the Freedom Tower.

MEANWHILE, AS I HAD been battling with the insurers, a design, despite all my misgivings, had been approved for the new Freedom Tower (as Governor Pataki insisted on calling what I, and most people for that matter, were simply referring to as One World Trade Center). And in the late fall of 2003, the construction crews had begun the preliminary work to excavate the site. This was painstaking, complicated, and expensive. First the damaged slurry wall along Vesey Street had to be braced to keep it from collapsing. Then heavy machines were brought in to pound through the deep bedrock of Manhattan schist so that the column footings anchoring the 1,776-foot-tall tower could be planted. And the job needed to be precise: the crews burrowing through walls of solid rock had to take care to avoid the snaking network of active commuter and subway train tracks that ran beneath the site. The main tower columns, in fact, needed to be built literally between the PATH tracks. Yet as if all this wasn't already a colossal challenge, the entire project was also a race against the clock. The crews had been given just nine months to get the job done.

Excavation takes place for the future One World Trade Center
seventy feet below the street.

Why this deadline? In a word—politics. Originally Governor Pataki had effusively promised that "as fireworks burst in the sky" the nation "will begin to reclaim our skyline with a permanent symbol of our freedom. On July 4, 2004, we will break ground on the Freedom Tower." But that fulsome vision needed to be quickly shoved to the side. It just didn't seem appropriate, the governor's critics raged, to have all this pomp and circumstance surrounding a ceremonial spade's digging into ground where so many had died on 9/11; it would be like desecrating a cemetery. So the chagrined governor retreated and came up with a display that would be more decorous than breaking ground. On July 4, 2004—almost three years after 9/11—there would be a public ceremony to lay the cornerstone that would signal the formal start of construction of the Freedom Tower.

In preparation for this carefully orchestrated Independence Day show, a five-and-a-half-foot-tall block of polished garnet-flecked granite—garnet being the official gem of New York State, the governor's office explained in a press release—had been shipped down to the site from a quarry in the upstate Adirondack Mountains. The massive cube weighed twenty tons, and a tall crane was needed to drop it gingerly into place for the event; eventually, it would be lodged belowground in the seventy-foot-deep foundation. Etched across the cornerstone's face, in tight block capitals, were soaring words meant to convey what the rising of the 1,776-foot-tall tower would convey to the world: "a tribute to the enduring spirit of freedom."

On July 4, a rousing ceremony was played out for the TV cameras. The governors of New York and New Jersey, as well as New York's mayor, spoke with passion about the rebirth of downtown Manhattan on a summer's day that also celebrated the birth of our republic. But as I stood near them, one of the five hundred or so invited guests, I found the reading of the Declaration of Independence by a thirteen-year-old boy to be the most moving part of the program. His father had been a Port Authority police officer who had died in the terrorist attacks, and the boy's stoic dignity broke my heart.

At the end of the day, as we all walked away from the ceremony, I could not help but have a sense that, despite all my concerns about the design of the new tower, we had reached the end of the beginning. We would soon be entering a new juncture in the nearly three-year battle to restore the site. At last the rebuilding could begin.

IN MAY 2005, HOWEVER, ten months after the cornerstone ceremony—and, even more infuriating, over a year after the original design for the Freedom Tower had been approved—just as the concrete and steel was scheduled to arrive, the New York Police Department sent the Port and me a multipage document stating that they had security concerns about the building. And, shrewdly, they made sure that *The New York Times* knew about this report, too.

Echoing much of what I'd originally pointed out, the report worried that the building was only twenty-five feet away from the heavily trafficked West Street section of the West Side Highway. Its location would make the tower extremely vulnerable to blasts from car bombs; Department of Defense standards recommend a one-hundred-foot distance from the street. And the building's open base, with its rows of exposed columns, as well as the off-center spire, the police pointed out, only made things worse: they were accessible targets for terrorists, accidents waiting to happen. After all, this was, they reminded everyone, a site that had been attacked twice previously. It required a meticulous security plan, arguably as stringent as any in the country.

Governor Pataki seemed furious. He was reportedly counting on this building to be his greatest legacy, the singular accomplishment that, after three terms as governor, would energize his presidential campaign. Adding to his ire, the location that was being criticized was the exact spot that had been shown for One World Trade Center on the master plan, which the city had approved a year earlier. He tore into the police department and the Bloomberg administration for waiting so long to share their concerns. But then, to the governor's further embarrassment, it came out in the press that back in

August 2004, the department had sent the Port a letter outlining the problems and requesting an immediate meeting. The Port, incredibly, said it never received the letter.

Lost in the mail? Was that true? Who knew? But what I did know was that the name-calling and the anger were irrelevant. And when Joe Seymour, the executive director of the Port, told me that he didn't have to listen to the New York police, that they didn't have jurisdiction over a state agency, I realized he just didn't get it. "You have no alternative," I lectured bluntly. "This is a fight you can't win."

I am in the business of constructing buildings that tenants will want to occupy, and it was abundantly clear to me that no one would want to lease space in a building where the police had security concerns, especially when it would be going up on a site that already had been targeted on two other occasions. And who, for that matter, would insure it? Or lend money for its construction? Yes, a redesign would cost millions in architects and the construction fees necessary to prepare a new site. And it would delay the project by months, maybe years. But I also knew that we didn't have a choice.

So I met with the governor. I told him, Sure, this is unfortunate. I moaned along with him that it was in fact terribly, terribly unfortunate. It shouldn't have happened, I wailed.

But it did, I said as I quickly switched gears. And now we have to work with what the police gave us. We must discard the old design and start from scratch. It is the only practical way—unless we want to put up a building that no one will want to enter.

All the time I was making this case, I was also very circumspect. I didn't tell him we needed a design that would be more like the one I had originally championed, one that wouldn't be encumbered by all of Libeskind's unrealistic (and potentially dangerous) architectural flourishes that had so caught Pataki's fancy.

So what did the governor do? Well, he was a politician, and the good ones are by nature pragmatic. He realized there would be no point in challenging the police department; it would be a battle he'd never win. So he decided, just as I had, that it was time to pivot—

only, to my surprise, he put a condition on his acquiescence. In the best of circumstances, working full speed, it would take an architect at least four or five months to redesign a complex office skyscraper. Pataki said it had to be done in eight weeks.

"IT'S AS IF YOU'D asked an architect to design a three-bedroom house," David Childs explained with apparent dismay as he set out to tackle the redesign, "and he'd finished the working drawings and started construction—and then you said, 'Oh, no, no, no. What I really want is to have a small hospital here.'"

And, sure, the challenge David faced was daunting, but I was elated he had been now given the role of principal architect. After the police department's objections, Pataki had come around to realizing what I had been saying all along: Libeskind, despite all his talent, just didn't understand the complexities involved in putting up a massive office tower in Lower Manhattan. And while Libeskind still had the official title of master planning architect, it was now firmly established that Childs, the man I had originally recruited to spruce up the Twin Towers and had hired to design the new Tower Seven that was already rising, would be running the show. Just as I had always wanted.

When David went to work on the redesign, he did a lot more than merely satisfy the security concerns. Yes, he moved the tower to the east, farther away from the dangers of the West Side Highway. And he shrank and squared the building's base into a two-hundred-foot steel-clad concrete cube that would be firmly anchored in solid bedrock, and at the same time a distance away from the vulnerable slurry wall and the maze of underground train tracks. It was a foundation that would withstand practically any blast; this tower would not come tumbling down like its predecessor.

But he also came up with a design that I felt would be, like New York's Chrysler and Empire State Buildings, whose classical lines it echoed, a landmark rising up in the city's majestic skyline. The elegant glass-clad building shot up from its bulky, fortified base like a glistening spear. And as it ascended high into the sky, its edges were sharpened and shaped to form shining multifaceted triangles. Then the beveled shapes, in a dramatic bit of visual magic, coalesced into an octagon, becoming smaller and even smaller as they approached the summit. The tower's lines were clear and timeless, a dazzling calculus of geometric design. On its top, a spire, anchored by a round and regal crown, rose up high into the sky. And this spire was centered, no longer the precarious spear that could potentially be hurled to the ground as in Libeskind's design.

I loved it. It was not just beautiful, it was inspiring. It was just the sort of iconic building to which tenants would flock. And David, pro that he is, finished his conceptual drawings by the governor's draconian deadline to boot.

After all the debates and delays, after all the money spent on master plans, on designs and redesigns, I felt construction could at last begin. And once again I was raring to go.

AND WHAT ABOUT THE cornerstone, the twenty-ton symbol of the rebirth of the World Trade Center site, that had been put in place with so much ceremony on July 4, 2004? Well, it had not initiated the start of construction as all the politicians had fervently insisted at the time. That had been a sham, nothing but a massive photo-op to keep themselves in the limelight. Now the massive block of granite just sat there, ignored and useless, the uplifting words etched into the dark stone covered by a blue tarpaulin.

And as the new design progressed, it became evident that the cornerstone's present location was obsolete. When the tower shifted about forty feet westward, the adjustment left the cornerstone standing outside the proposed building. The tower's reconfiguration required that it would need to be moved, too.

Rather than risk the cornerstone's being damaged as a new eastern footprint for the tower was excavated, it was decided it would be wiser to remove it from the site. On June 23, 2006—nearly two full years after it was put in place to mark the start of construction—it was hauled away to Hauppauge, New York. Only this time there was no ceremony. Nor, for that matter, did anyone point out that the construction had still not started in earnest.

After the massive granite block was driven off in the back of a flatbed truck, the Port tried to save face by announcing that it would be returned in a year. As things worked out, that was overly optimistic. And as the years stretched on, I began to feel that the fate of the cornerstone was in many dispiriting ways a grim metaphor. It had become a tangible symbol of the grandstanding politics and expensive miscalculations that characterized the entire government-driven rebuilding process in which I found myself enmeshed.

Larry Silverstein joins New York governor George Pataki,
Mayor Mike Bloomberg, and New Jersey governor Jim McGreevey
at the cornerstone laying ceremony on July 4, 2004.

TWELVE

YET DESPITE THE MANY delays, despite the infuriating battles with the insurers, despite the politicians' self-serving agendas, and despite the fortune in ground rent I was laying out each month for sites I still couldn't build on, here was the thing—I never lost my determination about what needed to be done. Or for what the site could one day become. I still managed to hold on to a vision of what these sixteen ruined acres would look like.

Now, I'm not claiming I had at this point a specific picture in my mind's eye of the towers that would be built; I'm a developer, not an architect. But I did have a guiding philosophy, a credo, if you will, that I rigorously applied as I juggled the thousands of decisions, both big and small, that had to be made as we moved forward: I wanted excellence.

And sure, part of this was vanity. In my career, I had put up many buildings and I had made many deals, but I knew that what was done at the World Trade Center site would be my legacy. It would be how I'd be remembered. When you are in your mid-seventies, you start to think about those sorts of somber things. I didn't want to be known as the villain who plunked down a few sterile buildings in Lower Manhattan that had all the character of cardboard boxes. I could still recall with a painful twinge what happened when Klara first walked into the lobby of the original Seven World Trade Center I had built. She took one glance at all those walls of Carmen Red granite from Finland and said, "It looks like a mausoleum." And she was right. Sure, that building was a very efficient moneymaker, but now I was determined to do something more consequential. I wanted to create something that would look spectacular.

And, to be perfectly candid, I also knew that the better a building's aesthetics, the better it would be for business. You didn't need to be a veteran developer to understand the obstacles I was facing: I wanted tenants to return to a site that had been targeted on two earlier occasions by terrorists, to a neighborhood whose air had turned toxic in the days after the Twin Towers had collapsed, and I also wanted them to pay top dollar for the offices they would be leasing. The only way to accomplish both of those goals at a time when lots of people were dismissive of anyone's ever returning to Ground Zero (including, incredibly, the Port Authority chairman, who said that under no circumstances would he require his employees to work in the Freedom Tower) would be to offer them something special. That is, buildings conceived by world-class architects, technologically perfect, and state of the art in terms of environmental considerations and energy consumption. My motivating commitment: the latest, the best, and, of course, totally safe and secure. It wouldn't just be good design. It would be good business.

But above all, there was another more important consideration that I kept fixed in my thoughts. It always pointed a clear way forward for me through the years of frustration: I was building a monument to those who had died on September 11. It had to be a fitting tribute; its towers had to possess a power and grace that would honor their memory. At the same time, what rose up from the ashes needed to bring life back to a part of Manhattan that had been destroyed. It had to be a spectacular complex that would attract tourists and help revitalize a dispirited neighborhood for families and office workers who had been deeply shaken by the terrible memories of an attack against our nation in their own backyard. It had to be something that would show the world that America doesn't give up, that we don't surrender. Instead, we rebuild, and this time we do it bigger and better. I felt that what I did on those sixteen acres should serve as a national landmark, a defiant response from the entire country to those who had thought they could destroy us.

IT WAS ESSENTIAL, THEN, for me to hire the best possible architect to design the three buildings I would be erecting. It had to be someone whose talent would lead him to envision soaring designs, and at the same time it had to be someone I was confident could tackle such a monumental job and succeed in a practical way, too. I needed an architect who would create office towers with the tenants in mind, skyscrapers that could be built for the real world on a realistic budget. And from the start I was certain I knew who it would be. I had only one person in mind.

David Childs had won me over. Before I had moved forward to buy the leasehold on the Twin Towers, one of my first decisions had been to bring David on board to reinvigorate the aging lobbies. Then, when I hurried to rebuild Tower Seven, I once again had turned to David. Keeping to a tight schedule, he not only came up with a stunning design, but also worked with me to restore the previously abandoned street grid. Like me, he understood that it was vital to turn downtown into a neighborhood, to attract people and families who would stroll through the Lower Manhattan streets. And now, a professional and a gentleman, he'd navigated a tricky (and often turbulent) collaboration with Libeskind, delivering a swiftly revised plan for the Freedom Tower that not only satisfied the police department's security concerns, but also promised a skyscraper that would become a classic, a glittering, sculptural addition to the New York skyline. I had no doubt that with David and his Skidmore team at the drawing board, I could get three more buildings that would help make my vision for the site a reality.

So with my mind made up, I offered him an opportunity that I was certain would take him by surprise as well as fill him with joy. It would allow him to stamp a singular vision on the city in a way no other architect had done since Raymond Hood had designed Rockefeller Center.

"How would you like," I asked, unable to disguise my own sense of excitement, "to design the three other towers I'll be constructing on the site?" T. J. Gottesdiener, Skidmore's managing partner, was present when I made the offer, and I could see he immediately grasped the treasure I was offering. He broke out in an exuberant grin. This would be, after all, not just a great opportunity for David, but for the entire firm. And it would be quite a windfall for Skidmore; the architecture fees for three skyscrapers would be enormous. Then I turned to David. I had no doubt he would quickly accept my offer.

Instead, David turned me down. I was surprised, but my reaction was nothing compared to Gottesdiener's. He looked shocked. His face suddenly turned very pale. For a moment, I wondered if he might collapse.

David, however, was undeterred. The very nature of New York, he thoughtfully began to explain, was its multiplicity of designs. It was a diversity that reflected all the many ways of life that kept the city churning away so creatively, and, he felt, it was an amalgam that helped to fuel the city's hard-driving energy, too. If the World Trade Center site was going to be transformed for a new generation, he firmly believed it shouldn't be stamped with a singular vision. It should, instead, be driven by the visual excitement several architects, each with their own way of looking at things, each conceiving a unique building, could bring to the neighborhood.

I listened, and when I recovered from my initial surprise and disappointment, I understood that he was right. The site would benefit from a multiplicity of visions, a roster of different talents. Like the city itself, the different parts would work together to create the whole. And I also appreciated that David was indeed a good friend, someone who wanted what was best for me, and the city, too, rather than simply for himself. People like that are very rare. I respect David immensely.

Only now I had to find the singular talents who could get the job done.

WITH DAVID HELPING TO pave the way, often making the initial introductions, Klara and I (sometimes with friends in tow) began to tour the world. We looked at buildings in Japan and across Europe, and we spent time with celebrated architects. It was an education in different cultures, different visual approaches and sensibilities, and distinctly different personalities. We met with architects with egos as large as the edifices they had designed, and with architects who possessed more restrained characters. And all the time I was guided by my desire to bring the best possible architects to the site: I was committed to excellence.

In the end, just as construction was starting on the Freedom Tower, I hired an all-star team of international architects, three Pritzker Prize winners. Fumihiko Maki with his exquisite, restrained fusion of Eastern sensibilities and Western innovation would design Tower Four. Adjacent to it on Greenwich Street, Richard Rogers, who saw things in a functional, often jarring, visible nuts-and-bolts sort of way, would design Tower Three. And farther up Greenwich, where it intersected with Dey Street, Lord Norman Foster, with his breathtaking high-tech modernism, would design the final remaining tower, Two.

But once the contracts were signed, I didn't just wave good-bye to the architects and their teams and send them back to their offices across the world. The plans and specifications for the three buildings were, according to the latest agreement I had negotiated with the Port, due on April 30, 2008. This was a very tight deadline, especially when you consider how huge an undertaking it was: the three towers would contain 10 million square feet of buildable space. And while these architects were extraordinarily accomplished, world-renowned, they also had egos to go along with their accomplishments. I couldn't afford for the architects and their teams to go off on tangents, to waste time. There was a deadline and I was determined to meet it. Mickey Kupperman came up with a brilliant idea.

Let me tell you about Mickey. I first met him in 1990 when Israeli prime minister Yitzhak Shamir asked me to help solve the housing problem caused by the sudden influx of 250,000 immigrants from Russia to the country. Mickey headed A. Epstein and Sons, a construction and design firm headquartered in Chicago, but also with offices in Israel and Poland. The company had built well-fortified secret air bases in Israel and thousands of units of low-cost housing in Sweden as well as elsewhere in Europe. As a result, Mickey knew a great deal about building the sort of reinforced concrete, blast-resistant low-cost housing the Israelis wanted for the new wave of immigrants.

As it happened, the Israeli government ultimately decided to use local developers for the project. But after all the time we had spent together, Mickey and I became friends.

And in the weeks after 9/11, I asked him to come to New York. I wanted to give him a tour through the ruins, but more importantly I wanted his advice on how we should proceed. I knew that Tower Seven, the first building to go up, would need to be a fortress, impervious to a terrorist attack. And Mickey had years of experience in building bomb-resistant structures. I asked him to work with me on the new construction.

"I'm retired," Mickey explained.

I kept asking, telling him how important the rebuild of the complex was to the entire nation.

"I can only give you two years," Mickey finally agreed.

Eighteen years later, Mickey is still working with me.

And back then, in the early years of our working together, it was his innovative idea that I quickly shared with the architects I'd hired to design my towers. "I'm setting up a design studio for you," I told them. "You're going to work from the tenth floor of my building right at the World Trade Center, number Seven. I'll give you everything you need. Computers, desks, drawing boards. Whatever's required, you tell me, and it'll be there. And you can have twenty-four-hour access to the studio. You want to work at midnight, be my guest."

They all jumped at the chance. Each of the three architectural teams had their own separate spaces in the tenth-floor studio, but at the same time it was a communal project. Every night the newly drawn plans for each of the three buildings would go up on a wall, and everyone could see the progress that was being made. And to emphasize that time was short, that there was a looming deadline, a calendar—again Mickey's idea—was prominently fixed to the wall, showing precisely how many days remained until the final plans would be due. At the start of each new day, a page would be ripped off, a very visible—and motivating—reminder that time was fleeting.

3 World Trade Center architects Rogers Stirk Harbour + Partners present designs to Larry Silverstein and his team, Mickey Kupperman, Janno Lieber, and Bill Dacunto.

Also, there was another distinct advantage in having the three teams working out of a shared office adjacent to the site. While each of the three towers would, of course, be its own unique vision, this communal experience gave the architects the opportunity to consider in very practical ways how each of the distinct structures would relate to the others. And by working in a building on the World Trade Center site, the designers each day had to think about how their individual towers would aesthetically and functionally interact with the rest of the redevelopment plans, including the 9/11 Memorial and the massive below-grade infrastructure of commuter train lines and parking garages the Port was constructing.

This unprecedented exercise in collaboration was a tremendous success. On April 30, 2008, we had the detailed plans for the three buildings just as the Port had requested. Three of the most celebrated architects in the world had delivered designs for three towers, 10 million square feet of space, and they had done it at the cost of $125 a foot. That is, I paid out a total of $125 million for the entire cost of the three designs.

Yes, that's a lot of money. But let me tell you, it was a bargain for what I got in return. Especially when you consider what the Port wound up paying for the design of a single building, the World Trade Center train station. They spent over $400 million to design an 800,000-square-foot building of walkable space. That's $500 a square foot for the architectural plans of one building, or four times what it cost me for the designs of three separate skyscrapers by three Pritzker Prize winners.

In fact, the story of the design and construction of the Oculus, as the downtown transit hub would become known, is a maddening tale of how the Port, working in tandem with many politicians who were driven by their own political motives, went about its business. It is a case study of what can happen when politicians play at being developers.

.   .   .

Completing the Vision. The first image of each of the new World Trade Center
towers presented to the public on September 7, 2006

BUT FIRST I NEED to make a confession. Before sharing my rage
over how the Oculus achieved, in even the judgment of the usually
restrained *New York Times,* its "ignominious distinction as one of the
most expensive and most delayed train stations ever built," I should,
in all fairness, acknowledge the small role I had played in the events.

I was the one who had first suggested to Governor Pataki that he hire Santiago Calatrava.

Klara and I had visited the Milwaukee Art Museum, which Calatrava had designed, and it was just spectacular. With stark white flying buttresses and ribbed vaults, the building was perched above Lake Michigan like a twenty-first-century interpretation of a Gothic cathedral—only it had a roof that opened at the press of a button. In fact, I was very eager to experience how it looked when the spires parted. Except there was a problem. The roof opened automatically on a preset schedule, and we had to leave before the appointed moment. But I coaxed and coaxed, and finally the museum people kindly relented; they would give us a special demonstration. And when the roof parted and a patch of blue sky appeared, well, I guess you could say I was hooked on the uniqueness of Calatrava's talent. The architecture was magnificent, totally inspiring. I had no doubt the man was a genius. (Later, Klara and I went to Spain to see more of his buildings, and we got to spend time with him and his family. Calatrava was a charming host, and the more of his work I saw, the more I was affected by the power of its sublime beauty.)

It wasn't long after I returned to New York from Milwaukee that I heard that Governor Pataki was searching for someone to design the PATH terminal at the Trade Center. But he wasn't looking for an architect who would come up with merely a well-functioning train station. He had grander aspirations for this transit hub—and for himself. His intention was to build the next Grand Central Terminal, a train station that would claim, as Pataki would boldly say, "its rightful place among New York City's most inspiring architectural icons." He also, according to what I had heard, had found the man he wanted to design this would-be masterwork—Daniel Libeskind.

Now, Libeskind and I had made our peace, and I respected his talent. But the prospect of Libeskind supervising the construction of a complex transit hub—fourteen train lines converged belowground—and getting it done on a schedule that would allow me to get my towers up and running as my agreement with the Port stipulated, well, that made me apprehensive. Pataki had already made one mistake by bringing Libeskind in to design Tower One, and now, I feared, he was preparing to compound it.

I told the governor about Calatrava. He was an architect, I said, who could design something monumental, a building people would talk about. And having learned after all my years in the real estate business that a good picture was often more persuasive than a lot of hyperbolic talk, after our conversation I sent over several books illustrated with Calatrava's work. I knew they would be the clincher; the full-color photos were breathtaking, the ingeniously conceived buildings inspirational.

But at the same time, I gave the governor a warning: Calatrava was expensive. Unbelievably expensive. His buildings cost a lot to put up and they cost a lot to maintain. They were the sort of structures only governments can afford. And when you hire a man like Calatrava, a person of extraordinary accomplishments—he was an engineer, a mathematician, a potter, and a sculptor, in addition to being an architect; the Metropolitan Museum had given him a retrospective in midcareer—you were also getting someone whose ego was proportionate to all the acclaim he had received. He wouldn't be an easy guy to keep in check. When you add that sort of showman's personality, that grandiosity (however well deserved), to the fact that his exquisite, inventive buildings could only be constructed at a phenomenal price, you had better be sure you could find someone who will keep a firm hand on the project. You would need to exercise control or else the cost will become astronomical. You would need a project manager who would stand up to a genius unwilling to compromise his expensive vision.

Did the governor hear the cautionary caveats I offered? I don't know; he never said anything substantive to me either at the time or afterward, when it was too late. All I know was that once he looked at the photos in the books I had sent him, Pataki was enthralled. He wanted Calatrava to design the terminal for the downtown transit hub.

. . .

WHAT THE GOVERNOR WANTED, the governor usually got. In 2003, the Port chose the Downtown Design Partnership, a joint venture, to be responsible for the plan for the entrance and aboveground terminal for the Trade Center Transportation Hub. The partnership decided on Calatrava as their architect.

A year later, he unveiled an ambitious vision. Its focal point was its main terminal, which he called the Oculus, a fancy architectural term that traditionally refers to the opening at the apex of a dome and that referenced his operable roof. As designed, it was a big space. The Oculus would be about twice the size of Grand Central Terminal's main concourse. And it was beautiful. The terminal rose up to a roof with two movable wings that could open to the sky; even when closed, the long, steel-ribbed glass wings were extended to the limits of the site, giving the structure the illusion of a bird in full flight. And it was grandiose. Perfectionist that he is, Calatrava had reportedly exceeded his specified mandate and designed more than just the passenger terminal; he included drawings for the concourses that would connect the main terminal to satellite train stations as well as plans for the train platforms and the underground mezzanine.

The Port reviewed the drawings and was bowled over by the prospect of such an iconic building. As for the additional elements that Calatrava had now included, after the architect insisted they were essential to the integrity of the project, the charmed (or was it cowed?) Port acquiesced to his designing them, too.

In 2004, the Port authorized a staggering $2 billion for "the Calatrava," as the project had become widely branded. The Federal Transit Administration would cover $1.7 billion of the cost and the Port would kick in an additional $300 million. And, of course, all of this $2 billion would be coming from the taxpayers.

But even before the project could get off the ground, there were concerns. First, the police department raised security issues. To ensure that the Oculus would withstand a terrorist blast, a major redesign,

Santiago Calatrava's concept sketch for the Oculus,
inspired by a dove taking flight

they insisted, was required. To Calatrava's protests, the graceful steel ribs were thickened and their number doubled. And as for the translucent glass in the "wings" that would flood the structure with natural light, this, too, was eliminated. *The New York Times* studied the new design and sneered that "it may now evoke a slender stegosaurus more than it does a bird."

In addition, as the redesign proceeded, Port officials began to have second thoughts about the Oculus's operable winged roof. Not only would it be exorbitantly expensive, they complained, but they also wondered if it was even feasible. So Calatrava, who possessed a showman's infectious charm, organized a trip to the Milwaukee Art Museum for Port officials and board members. As they stood

Architectural rendering of Santiago Calatrava's Transportation Hub,
known as the Oculus

outside the museum waiting for the roof to open, they were joined by a group of schoolchildren. When the spires parted to let in an expanse of sky, the delighted children, as if on cue, broke out into applause. It was at that rapturous moment, the applause echoing, that a board member realized he was waging a battle he'd never have the tenacity to win. "Okay, Santiago," he said according to the dismayed account reported in the *Times*. "You can have your goddamn wings."

And so in September 2005, the groundbreaking ceremony signaling the start of construction of the Oculus occurred. As at all the other carefully staged events associated with the Trade Center, the television cameras captured a crowd of politicians—the transportation secretary, four United States senators, two governors, and the mayor of New York were all there. Then with the cameras rolling, an ebullient Calatrava and his daughter, Sofia, released white doves into the downtown sky, symbols of the $2 billion building inspired by birds in flight whose roof would open dramatically. It would be completed, the governor pledged, in 2009.

The ceremony, as things turned out, was a sham. The Oculus would not be finished for more than another decade. The cost would double to a mind-boggling $4 billion. The gaudy plan for the operable roof would be abandoned. And as for the exotic doves, it was later revealed that they were run-of-the-mill pigeons.

IN MANY UNFORTUNATE WAYS, the construction of the Oculus exemplified what I was caught up in as I tried to navigate my partnership with the public sector to rebuild the Trade Center. My efforts, too, were sabotaged by politicians who had their own agendas and by Port Authority executives who didn't have the expertise to manage their contractors. And in the end, just as with all the other governmental screwups at the site, it was the taxpayers who got stuck with the bills for this folly.

It was Pataki, the would-be president, I suspected, determined not to anger the largely Republican commuters getting off the ferry from

Staten Island, who refused to close temporarily even part of the No. 1 subway line that ran through the site. This seemingly blatantly political decision forced the Port to build around and over the subway line—at an additional cost, according to newspaper reports, of $355 million. And it would delay my crews for years from putting in the initial foundations of the towers I would eventually build.

Then there was Bloomberg, who set a September 11, 2011, deadline for the completion of the National September 11 Memorial at the site. Once he announced the date, he was determined that nothing was going to stop him from having this moment. This timetable meant millions had to be spent on re-engineering the transit hub's roof to support the extensive landscaping the Bloomberg administration wanted surrounding the memorial. And since the construction of the memorial's deck was now proceeding simultaneously with the building of the Oculus's mezzanine level, the cranes that were already in place for the transit hub could not be used as originally planned to lower the necessary equipment and material. Instead, the Port had to buy ten flatcars that would run through the PATH lines as a freight railroad to deliver the material to the Oculus—for an additional $3 million.

But these excesses were, in the costly scheme of things, only rounding errors. The real culprit in this misuse of public monies, it seemed to me, was the Port Authority. Their managers were in way over their heads. Sure, they were hardworking. But that was not enough. They just didn't have the expertise to deal with rapacious contractors, nor the spine to stand up to a celebrated architect who persistently argued for the integrity of his unique vision. The sad truth, I came to understand, was that anybody of any quality or experience wouldn't be working supervising construction for the government. They would go into the private sector, where they would get paid a lot more generously for the work—and where, if they wanted to keep their jobs, they would need to keep a sharp watch on how every dollar was spent. In government, people had a tendency to shrug off cost overruns; it wasn't their own money they were spending.

For example, Calatrava insisted on curvilinear steel elements that could only, as it turned out, be manufactured abroad. That resulted in the steel bill for the entire project rising to an exorbitant $474 million. And when the Spanish steelmakers got behind schedule, the Port simply agreed to pay another $24 million for overtime. Then there was the contractor who wound up being dismissed midway through the project, but was still paid $982.5 million for only a partial share of the work. And even though the Port had its own team of managers in place, the Design Partnership received $405.8 million for its additional supervision, with about $80 million of that going to Calatrava's firm.

As I watched this free-spending disaster taking shape, I decided I should try to intervene. The constant delays, the staggering cost—I feared it would all spill over to affect the progress of the towers I wanted to construct. To be completely truthful, I was embarrassed, as well as enraged, by the cavalier way things were proceeding. There was a right way and a wrong way to get things done, and this was definitely the wrong way. After a half century or so in the real estate business, I knew I could help the Port.

So in the summer of 2008, I went to Chris Ward, who had just been appointed executive director of the Port Authority by the new governor, David Paterson (who replaced Eliot Spitzer, after he had resigned in disgrace). "The cost of this terminal is going crazy, absolutely crazy," I said. "Let my firm help you figure out ways that'll allow you to cut the costs. Maybe we can eliminate some of the complexities in the design, suggest some things that'll save you time and money. It's what we know how to do."

Chris, to his credit, jumped at the idea. Come on in and help, he agreed.

I had my people go to work. These were people who had been doing this for years, professionals who knew how to keep an eye on every penny because they knew they wouldn't have a job with me for long if they didn't. They worked out a plan that would save a huge amount of money and maybe a year or more of construc-

tion time. And, sure, we recommended some slight changes to Calatrava's design. But these were the sort of things most people would
never focus on in a million years. The terminal plans, for example,
called for a huge column-free space, and obviously that sort of design
required feats of engineering that are wildly expensive. If you put in
a supporting column, was it really such an eyesore in the scheme of
things? Especially when you inserted the column where most people
would never notice it? It wouldn't have had any significant effect on
Calatrava's design. The key for people was to get to or from their
trains as quickly as possible. They didn't care about the columns, the
design of the subway platforms. But if you eliminated some of the
unnecessary frills, it would save a ton of money and allow the terminal to be up and running sooner.

We shared these plans with the Port officials supervising the project, and they were clapping us on the back, telling us that what we
had done was fantastic. Then the next thing we heard was that the
Port was totally rejecting everything we had recommended.

What had happened? What had caused such a complete about-
face from their initial enthusiasm? I don't know. But I have my sus-
picions. I heard that Calatrava convinced a group of decision-makers
on the Port's board to come visit him in Valencia, Spain. He showed
them the magnificent buildings he had designed. The City of Arts
and Sciences and the Opera House—each spectacular works of art,
each iconic. By the time they had returned, they were drinking the
Calatrava Kool-Aid. They wouldn't hear of any changes to his design
even though my people had estimated that our suggestions would
save about $500 million. That kind of savings meant little to them.
They didn't need to keep to a budget. They didn't need to answer to
anyone for their spending.

WHEN THE OCULUS FINALLY opened in March 2016, neither the politicians nor the television cameras were invited. There was, in fact, no
ceremony, no fanfare. The executive director of the Port, who had

been appointed years after the project had gotten underway, issued a terse statement saying he was "passing" on holding an event. However, when pressed, he explained his decision more fully. The terminal, he said, had become a "symbol of excess."

But what did the taxpayers get for the $4 billion they'd shelled out? Here's David Childs's assessment: "There's no purpose to the building. It's a receiving hall. You can't get a ticket there, you can't get a Coke there—it's a vestibule. And any possible way that you calculate it, it's the most expensive building ever built by mankind. And it has no purpose."

And what did I think of the finished product? Well, even though Calatrava's original soaring, avian design had been scaled back and modified until it barely resembled the initial proposal, the building was still monumental, a work of art. But it also was a terminal that was twice the size of Grand Central, and it would be used each day by only a third as many commuters. And it had cost a fortune.

What the Port had demonstrated in its supervision and construction of the Trade Center transport hub was not just incompetence. It was gross incompetence. And it was the sort of capricious and cavalier behavior I had to deal with the entire time as I struggled with the Port to rebuild the Trade Center site.

THIRTEEN

T HE REBUILDING OF THE sixteen-acre complex was a complicated, many-faceted process, and so it was often the case
that many dramas, both public and private, unfolded simultaneously. Therefore, I want to share something else that was happening in 2005 as the preliminary designs of the Freedom Tower and the
Oculus were nearing completion and my newly constructed Tower
Seven was anticipating opening to tenants: the Port had decided it
was time to push me out of the picture once and for all.

This wasn't their first attempt. It seemed that from the moment
the rubble had started to be cleared from Ground Zero, the Port
had been eager to shove me out of the reconstruction process. They
claimed that my company was too small for such a massive undertaking, that I would lose interest and wind up unwilling to make
the commitment of years and energy that would be required, that I
would pocket the insurance proceeds and run.

What was different about their latest attack, though, was that they
had the politicians cheering them on like a Greek chorus. There was
Pataki, a lame-duck governor who was willing to do, or say, what-

ever he thought would help his cause in the presidential primaries. He was now arguing that even with the insurance proceeds and the government's Liberty Bonds, I wouldn't have the funds to get the entire job of reconstruction done. Mayor Bloomberg, too, was piling on, making me the villain. His deputy mayor Dan Doctoroff testified to the City Council that at most I would build only two of the five towers before running out of construction financing. Then I would simply default on my lease with the Port and steam off on my boat with hundreds of millions of dollars stashed away from the insurance proceeds.

Accusations like these stirred up the press. A raging *Times* editorial lectured that "unless Mr. Silverstein wants to be remembered as the man who ruined the legacy of the World Trade Center, he needs to reduce his demands." The *Daily News,* as I have previously pointed out, didn't even offer me the possibility of playing a role in the rebuilding. "Get Lost, Larry," it had summarily ordered in a large, bold headline.

Yet while the Port and the politicians had only managed to hurl venomous accusations or show up for photo ops at Ground Zero, I had actually done something: I had rebuilt Tower Seven. And I had been paying out $120 million a year in ground rent for the past four years—nearly a half-billion dollars!—and had spent hundreds of millions more on design costs and lawyers for the site. It was really very infuriating. At times I couldn't help thinking of the old adage that no good deed goes unpunished.

But then the governor, in December 2005, decided the time had come to break through the stalemate between the Port, which owned the Trade Center land, and Silverstein Properties, which had the lease. He gave us ninety days—till March 14—to come to an agreement that would allow construction to start in earnest on the site.

IT WAS AN ULTIMATUM that caused me to take a deep breath, and to begin to think about things differently. With the clock ticking, I had little choice but to consider my predicament in a more practical way: I couldn't build without the Port, and they couldn't build without me. Therefore, we needed to work together. We needed to talk respectfully to each other. As for all the name-calling, I had known it'd be coming. But that was business in New York. That's what happened when you butted heads against powerful interests in this city. I couldn't allow it to have any permanent effect, to run me off. Too much was at stake.

And if the only way to get it done was to sit down with the Port and work things through, then that's what we had to do.

When I thought about it all this way, when I allowed myself to get over the annoying fact that I was being shoved into a corner, I had to concede that the insurance proceeds would only be a small down payment on the construction costs. The entire project—five individual skyscrapers—would run to $20 billion. At least. And with a businessman's calm practicality, I started to think that maybe the Port was right: I should relinquish part of my development rights. Maybe, I decided, there was a deal that could be made.

BUT DID I REALLY want to give up my rights to the Freedom Tower? Once the Port and I had started talking, we batted around various compromises. Multimillion-dollar building sites were traded back and forth between us as if they were properties in a game of Monopoly. At one point, they offered me Site Five and $50 million simply to walk away; it was as if they had never heard me say I was never going to quit and run. Stymied, they moved on to suggest a totally different sort of plan: the Port and Silverstein Properties would jointly build and operate the entire new World Trade Center complex. But in the end, the Port and the governor, against my practical advice, wanted the 1,776-foot Freedom Tower to be the first building to be completed. And to satisfy their own political agendas, they wanted to be sure they would be the ones who'd get the credit for doing it.

This was a sticking point for me. I had already spent $150 million on the original drawings and schematics for the building, and then had shelled out another hundred million or so after the police department insisted the tower had to be redesigned. I had negotiated the building schedules with Tishman Construction, and then renegotiated those, too, after the location had to be moved for security reasons. And, not least, this building had been my baby. I had wrestled control out of Libeskind's hands to make sure that David Childs was running the show. And I had consulted with David constantly, supervising his many thoughtful revisions, as he worked to come up with a building that would dominate the downtown skyline and could also be built and then operated as a profitable commercial enterprise. I had been involved at the beginning, and I wanted to be there when tenants moved in. It was, I confess, a very emotional issue for me. I had put my heart into this building.

And that, I finally came to realize, was the problem.

I needed to start thinking, instead, with my head. And when I looked at the Freedom Tower with a developer's cooler objectivity, there were many good reasons to walk away. For one thing, it would be ridiculously expensive to build; the cost for the 3.5-million-square-foot building was projected to be about $4 billion. Rents would have to be set uncompetitively high to make a profit after that large an investment. But could it even attract tenants? The tower would stand on the northwestern corner of the site, about as far as possible from the subway and PATH line that converged into the transit hub. And, arguably a greater obstacle, would people want to work in the building that was the symbolic replacement for the Twin Towers? That might very well cause some reluctance, and the governor's insistence on calling it the Freedom Tower had always seemed to me a foolish gesture; as I had complained earlier, it was almost as if he had painted a bull's-eye on the façade, daring terrorists to attack.

In a world where the decisions were being made by a developer, not the politicians, Towers Two, Three, and Four were the best properties, and should be the first to be built. Not only were they right on Greenwich Street and therefore closest to the thriving neighborhood business district and the subway lines, but they also had larger footprints. Tower Two, in fact, had a footprint of about 70,000 square feet. That meant you could build a straight-shaft tower with a huge base and podium, the lower floors perfect for the sort of open-space trading floors the financial firms were all demanding in 2006. I could see those gigantic spaces renting quickly. The Freedom Tower, in contrast, offered only the possibilities of smaller floors; its floors started at 40,000 square feet and kept shrinking as the building rose, culminating in spaces only in the high thirties at the top.

The more I looked for a way to resolve my stalemate with the Port, the more I started to see the prospects for a deal. It would be a deal that could work because it offered something for both sides. And that was the fundamental principle, as I always said, for reaching agreements: both sides had to feel like they were getting something they wanted.

And so as the governor's deadline approached, the basic outline for a new arrangement emerged. We would keep 60 percent of the site to build on; the Port would have the remaining 40 percent. That is, Silverstein Properties retained the right to develop three office towers right in the heart of the downtown neighborhood, along Church Street between Vesey and Liberty Streets, while the Port would build the huge 3.5-million-square-foot Freedom Tower, as well as have the right to develop Site Five, the former Deutsche Bank Building, as both a commercial and residential tower.

This was genuine progress; the Port and I had come very far in a matter of months. But there was still a long way to go. Our broad strokes of an agreement left a myriad of other complex financial details that remained to be worked out. For example, how would the insurance proceeds be shared under this new arrangement? Would I get a reduction in the monthly ground rent I had been paying? And would I be reimbursed for the costly development work I had already done on Tower One, the Freedom Tower?

As the days counted down to the governor's deadline, I began to fear that the remaining issues would prove to be insurmountable.

What would happen then? Would the Port and I be fighting it out in the courts for years to come while Ground Zero remained a vacant hole?

YET BY THE MORNING of March 14, 2006, it looked as if, miraculously, everything would fall into place. My team sat down with the Port officials and Pataki's chief of staff, John Cahill, at the Port's headquarters on Park Avenue, and both sides went at it all day. By dinnertime, only about $50 million (in a deal involving billions) and a few minor issues such as the revenue split from the retail stores remained. So my team returned to my apartment, where I had been having discussions with two of my financial backers; this was, after all, very much their deal, too, and they'd have to live with its terms. We had dinner—I had ordered a feast from DB Bistro on Forty-Fourth Street, already in my mind celebrating—and at the same time we discussed strategies for putting the final touches to the deal.

When my team left for the Port's office, I returned with them. We must have arrived at about nine thirty, or maybe closer to ten p.m. It was past my usual bedtime, but I felt it was important to get the final issues ironed out. And I also wanted to be there so that I could directly communicate my positions; I didn't want the Port to assert later that they had misunderstood what was being proposed. So I drank a cup of coffee and waited for the Port officials to come into the fifteenth-floor conference room.

An hour later I was still waiting. They didn't show up till after eleven p.m. And they were seething. They said we had arrived after the dinner break later than we had promised and, I concede, there was some truth to that. I apologized and explained that I had needed to persuade my money people, who had been having second thoughts. This explanation failed to calm their mood. Ken Ringler, the executive director of the Port Authority, later complained that we had been "partying." We had food, he railed, from "a French restaurant while my people sat at their desks eating pizza."

And then things went from bad to worse. I simply asked for clarification on a few points, and all at once the Port officials exploded. They started shouting and cursing, claiming I was backtracking on several issues on which we had already agreed. To my mind, that wasn't the case at all. The way I saw things, the Port was once again in over their heads. I make real estate deals for a living, and the devil is always in the details. It was these fine points that needed to be worked out. That was how a deal was concluded. You dot all the i's and you make sure that all the deal points are crystal clear.

The Port, however, just didn't have the ability to negotiate a deal of this size or this complexity. And the fact that in addition to the representatives in the room, the governors of New York and New Jersey, as well as the Port's directors, all had to sign off on nearly every deal point, well, that sort of cumbersome machinery had made it difficult to work out the fine points. When I kept pressing for clarity, they were overwhelmed. Rather than try to think their way through the thorny issues, they threw up their hands in despair, let loose with a volley of expletives, and then announced that the negotiations were over.

While they didn't know about deal-making or real estate development, the Port officials did know about handling the press. They reached out to reporters and fed them their side of the story, and I took a genuine beating in the days that followed. Ken Ringler claimed that I had assumed I had leverage because the governor's deadline was approaching. I was using the ticking clock to force them to "back down" to my demands. And, more damaging, albeit inaccurate in my opinion, he asserted that I had tried to change significant deal points, items that had been previously agreed upon. "We are not going to make a short-sighted deal with Larry Silverstein," he raged indignantly at a press conference.

And the editorial writers believed him. I was castigated in the *Times* (and not for the first time, of course). A March 17 editorial was a one-sided, highly inaccurate version of what had happened. "The terms were overly generous to Mr. Silverstein, and he was very lucky to get them," the *Times* judged. "But at the last minute, Mr. Silverstein and his team made a new set of demands that seem to have been intended to scuttle the bargaining."

I held a press conference the next day to explain my side. But here's another truth I had learned since I entered the World Trade Center battles: a wealthy developer never gets much sympathy—forget about support—from the New York press. In the reporting of the breakdown of the negotiations, the narrative of events had been predetermined by the press, and the facts were irrelevant. In this war, too, the truth had been a casualty.

Yet did this mean the prospects for an agreement were finished? That the fight between me and the Port over who should develop the World Trade Center site would end up in the courtroom? I didn't think so. Part of my reason was my persistent optimism; it was in my nature to think things would work out. But I also knew enough about deals to understand that when something was on the table that offered both sides a win, there was a good chance issues would ultimately be resolved.

NEVERTHELESS, I WAS TAKEN by surprise when the Port announced their terms for proceeding with the negotiations: They refused to talk to me, or for that matter any of my in-house team. Instead, they'd only meet with the Wachtell lawyers whom I'd hired.

This was highly unprofessional. But when we were this close to dotting all the final i's, it seemed counterproductive to focus on the Port's childish behavior. Instead, I told the lawyers, "Guys, let's close the damn deal." But I also made sure that they understood they were not to close anything, not a single deal point, until they ran it by me and I signed off.

In this manner, a bit more disjointed, perhaps, than most negotiations, things began to be resolved. And on the morning of April 26, six intense weeks after the Port had pretty much thrown us out of their conference room, a conceptual framework for a deal was signed.

It was only twelve pages. The actual Master Development Agreement would run to more than two hundred separate agreements, would fill seven thick binders with intricate lawyerly sentences, and would not be completed until five months later. But this brief initial document gave both the Port and me a clear way to move forward with rebuilding.

I had made significant concessions. The Port would receive the Freedom Tower (although I would remain a consultant on the project). I would hand over about $970 million of the proceeds I had recovered from the insurance on the Twin Towers, and this would be used, along with $1.7 billion in tax-free Liberty Bonds, to fund the Port's construction of the new tower. In addition, I would contribute $140 million in common infrastructure costs. As I said, these were real—and expensive—sacrifices.

And what did I get in return? I received the right to build three towers on Greenwich Street. Each would be nearly as tall as the Empire State Building. In total, the buildings would have 6.2 million square feet of office space, and to ease the rental challenges, the Port and New York City agreed to lease 1.2 million square feet—about one-sixth of the entire space—at market rates. Also, to fund the construction of the towers, I would receive access to $2.6 billion in tax-free bonds the federal government had issued to encourage construction in Lower Manhattan. In addition, I would be building the Freedom Tower for the Port, and for that I'd receive a 2.5 percent development fee, a sum that would work out to be about $20 million.

Was it a good deal for me? I was quoted at the time as saying, "I'm not complaining." And I'm still not.

But there was also another aspect to the deal. The politicians had been worried that I might not finish my towers, that I would simply take my insurance money and run. I found the implications insulting, but I allowed them to insert what they considered pretty onerous terms into the agreement to satisfy their concerns: I would need to start my work on all three towers simultaneously, the construction to begin as soon as the Port finished digging out and fortifying the "bathtub" on the property and completing the necessary work for the belowground infrastructure, the subways and rail lines, as well as the underground tunnel network. The deadline I was given was December 31, 2007, for Towers Three and Four, and June 30, 2008, for Tower Two. And if I hadn't begun construction in earnest, there were not just very steep financial penalties; the Port could reclaim the sites. All three of them.

Sure, putting up three huge towers at the same time would be a challenge, but it was one I was looking forward to. I wanted to get this site rebuilt, and I wanted to get things moving as soon as possible. That had been my ambition since September 12, 2001. So against the advice of my team and lawyers, I accepted the terms. It was a crazy, stupid risk, one where I could lose everything if I didn't build on their schedule. Nevertheless, I took it. After all the years of delay, I was raring to get going.

But prior to my signing, I decided I needed to put some pressure on the Port to live up to their commitments, too. Therefore, I had another provision inserted into the agreement. In case the Port failed to get the belowground infrastructure work done within the time frame they had specified, then they would have to pay penalties. After all, they would be holding me back; I couldn't begin construction until they completed everything that needed to be done below grade.

But what should this penalty be? I gave this a good deal of thought. I worked out that I had been paying the Port about $300,000 a day in ground rent for the past seven years. If that was good enough for them, it would be good enough for me, too. I would require that they fork over $300,000 for each day they kept me waiting, unable to begin construction, unable to fulfill the timetable they had set for building my towers. That would be fair, I decided, and the Port agreed.

Still, at the time when the daily penalties were inserted into the deal, as well as another provision that required arbitration rather than endless litigation as the means for us to settle any potential

At the time of their completion in 1973, One World Trade Center (the North Tower) and Two World Trade Center (the South Tower) were the tallest buildings in the world at 1,368 and 1,362 feet, respectively. Other buildings in the complex included a hotel (Three World Trade Center); Four, Five, and Six World Trade Center; and Seven World Trade Center, opened by Larry Silverstein in 1987.

Annette, Etta, Harry, and Larry Silverstein at a formal dinner in 1947. The family lived in Bedford-Stuyvesant, Brooklyn, and later moved to Washington Heights in Upper Manhattan.

Larry and Annette Silverstein at Larry's NYU graduation in 1952. Larry founded the NYU Real Estate Institute in 1967, and has served as a member of the NYU Board of Trustees for more than sixty years. He is also a member of the NYU Langone Health Board of Trustees.

Silverstein on top of world with WTC deal

By ERIC HERMAN
DAILY NEWS BUSINESS WRITER

With several strokes of a pen, developer Larry Silverstein became the city's largest commercial landlord yesterday.

After weeks of negotiation, Silverstein took control of the gleaming 110-story Twin Towers of the World Trade Center, with Port Authority officials handing over the complex for 99 years.

Silverstein Properties will run about 10 million square feet of office space contained in the Twin Towers, plus World Trade Center buildings 4 and 5. Westfield America, Silverstein's partner in the deal, will operate the vast retail mall, most of which is below ground.

The deal represents a triumph for the 70-year-old Silverstein, who competed against some of New York's most powerful developers and real estate companies. During the past several months, his company survived several rounds of bidding and the initial selection of another developer — Vornado Realty — as the winning bidder. Port Authority talks with Vornado later broke down, and the agency turned to Silverstein.

"This is obviously a very humbling experience and one of high emotion for me," Silverstein said yesterday at a public ceremony commemorating the handover.

Under the terms of the deal, Silverstein and his partners will pocket rents from the WTC's office and retail tenants. In return, they will pay the Port Authority $116 million a year for the first five years, $138 million a year for the next five years, and higher payments after that. The payments begin in August.

In addition, Silverstein and Westfield made an up-front payment of $616 million yesterday.

The Port Authority has valued the deal at $3.2 billion. While the total dollar value of the yearly payments exceeds that price tag, the agency assigns less value to future payments for accounting purposes. Authority officials declined to specify the total dollar value of all payments.

By putting the massive facility into private hands, Port Authority officials said it would be able to concentrate on its original mission of running transportation hubs while locking in a steady flow of funds from the buildings.

"We're insulating the Port Authority from the vagaries of the real estate market, because we're not supposed to be in the real estate market," said Neil Levin, the agency's executive director.

But the deal's value to the Port Authority could diminish if the city makes good on its threat to yank a key tax break. Currently, the agency makes an annual payment in lieu of taxes of about $25 million. But city officials assert that the 99-year lease is effectively a sale and ends the tax deal. City officials estimated that full property taxes on the buildings would amount to over $100 million a year.

"Our position is that if it's privatized, the city should be able to collect the full amount of the property taxes on it," Mayor Giuliani said yesterday.

Silverstein will cover the current tax payments, but the Port Authority will cover increases in tax bills, said an agency source.

with Lisa Colangelo

BIG DEAL, BIG KEYS Larry Silverstein, head of Silverstein Properties, holds up symbolic keys in 99-year deal to take over World Trade Center Twin Towers and buildings 4 and 5.

BUDD WILLIAMS DAILY NEWS

On July 24, 2001, Larry Silverstein completed the largest real estate transaction in New York history when he signed a ninety-nine-year lease on the 10.6-million-square-foot World Trade Center for $3.2 billion, only to see it destroyed in terrorist attacks seven weeks later on September 11, 2001. He has spent the last twenty-three years rebuilding the office component of the World Trade Center site, a $26 billion project that will have consumed the balance of his life. From left: Roger, Lisa, Klara, and Larry Silverstein and Leonard Boxer.

The Twin Towers were destroyed by terrorists on September 11, 2001. Seven World Trade Center collapsed later that afternoon.

2,753 people died at the World Trade Center on 9/11, including 343 firefighters. It was the deadliest terrorist attack on U.S. soil and had a profound and lasting impact on the country.

The topping-out ceremony for the new 7 World Trade Center on October 21, 2004. Five hundred union construction workers celebrated the milestone with Larry Silverstein, Governor George Pataki, and Mayor Mike Bloomberg.

The ribbon-cutting ceremony for 7 World Trade Center on May 23, 2006, in the new park in front of the building. From left: Jeff Koons, Larry Silverstein, Dan Tishman, Ken Ringler, and David Childs.

7 World Trade Center was the first office tower to rise at Ground Zero after 9/11. The building was designed by David Childs of Skidmore, Owings & Merrill, and was the first LEED-certified office building in New York City.

In 2002, the Lower Manhattan Development Corporation (LMDC) held a design competition for a master plan for the sixteen acres destroyed by the terrorist attacks. In December 2002, Foster and Partners was among six finalists from more than four hundred international entries.

Daniel Libeskind's design, Memory Foundations, won the competition. In designing the master site plan, Libeskind worked closely with all the stakeholders, balancing the memory of the tragedy with the need to foster a vibrant and working neighborhood.

On September 7, 2006, Larry Silverstein unveiled the plans for 2, 3, and 4 World Trade Center, designed by Pritzker Prize–winning architects Fumihiko Maki, Norman Foster, and Richard Rogers.

Larry is joined in a design meeting by Janno Lieber, president of World Trade Center Properties, and Mickey Kupperman, the company's chief operating officer. Janno oversaw much of the World Trade Center rebuilding effort for fourteen years.

Larry Silverstein looks out over the empty World Trade Center site from the new 7 World Trade Center in December 2007. Work had begun on One World Trade Center, but the Port Authority struggled to excavate the land to the east of Greenwich Street—the sites for Silverstein's 2, 3, and 4 World Trade Center, and the Transportation Hub.

Union construction workers rally outside 7 World Trade Center in March 2010 to persuade the Port Authority to support Silverstein's efforts to build 3 World Trade Center.

Every year around 9/11, Silverstein hosted a press conference with key World Trade Center stakeholders to update the media on progress at the site. From left: Michael Arad, Joe Daniels, Sheldon Silver, Mike Bloomberg, Chris Ward, Daniel Libeskind, and Larry Silverstein on September 7, 2011.

Tens of thousands of union construction workers built the new World Trade Center over the past twenty years. Many were women and minorities, and apprentices included high school seniors and veterans returning from Iraq and Afghanistan.

Crane inspectors at work on the ninetieth floor of One World Trade Center the morning after the building was hit by a lightning strike. This photo by Joe Woolhead was published around the world and came to symbolize the dedication of the men and women rebuilding the World Trade Center.

The National 9/11 Memorial park takes up half of the sixteen-acre World Trade Center site. The memorial opened on September 11, 2011, the tenth anniversary of the terrorist attacks.

The topping-out ceremony for 4 World Trade Center on June 25, 2012. Built by over three thousand union workers, 4 World Trade Center was the first office building completed on the historic sixteen-acre site.

President Barack Obama speaks at the dedication of the National 9/11 Museum on May 15, 2014. Designed by Snøhetta, the museum tells the history of the terrorist attacks.

The 3 World Trade Center ribbon-cutting ceremony on June 11, 2018. From left: Gary LaBarbera, Carolyn Maloney, Kevin O'Toole, Rick Cotton, Marty Burger, Tal Kerret, Lisa Silverstein, Roger Silverstein, Larry Silverstein, Richard Paul, and Kelly Clark.

Lisa, Larry, and Klara Silverstein at the opening of 3 World Trade Center. Designed by Richard Rogers, the building was quickly leased to GroupM, Diageo, McKinsey, Uber, and other leading companies.

3 and 4 World Trade Center were designed by Pritzker Prize–winning architects Richard Rogers and Fumihiko Maki, and developed and opened by Silverstein Properties in 2013 and 2018, respectively.

The view from 7 World Trade Center looking south over the World Trade Center site and New York Harbor. This photo was taken on September 11, 2017, with the twin beams of light commemorating the attacks.

The new Perelman Performing Arts Center flanked by the Oculus, One and 7 World Trade Center, across the street from the site of the future 2 World Trade Center.

disputes, I had never thought they would become important. I never thought they'd have real consequences.

I WAS READY TO get to work. With the new Master Development Agreement (MDA) signed, I could at last begin the heavy construction of the belowground foundations for the three large towers along Greenwich Street I had just committed to build. And the timing sure seemed serendipitous—for both the Port and for me.

In June 2006, David Childs had delivered the final design for the Freedom Tower; this go-around he had covered the massive, bomb-resistant 186-foot-tall podium with a reflective screen of glass prisms that certainly looked a lot better than the brutalist steel and concrete in the previous plan. In Luxembourg, a foundry had started production on the 805 tons of steel—including the massive I-shaped columns—that would be shipped across the Atlantic to serve as the belowground anchors for the tower. And so in December 2006, just as the ink was drying on the MDA, there was still another carefully staged media event with Governor Pataki and Mayor Bloomberg once again front and center as a tall red crane lowered into place the first ceremonial steel column for One World Trade Center.

As for me, things seemed to be falling into place, too. My Tower Seven was not only up and running, but Moody's, a preeminent financial firm, had signed a twenty-year lease to occupy fifteen floors. It was tangible vindication of my unwavering faith that if you build it, they will come to the World Trade Center neighborhood—and pay a premium, to boot. Even Mayor Bloomberg, who had publicly doubted I would get the rents I was asking, graciously called to admit he had been wrong. And in addition, I was poised to finally receive about $4.68 billion from the long-gestating insurance cases; I had agreed to give the Port $970 million of this money, but from the remainder I would have the funds to pay for the preliminary below-grade construction work as well as my monthly ground lease rents to the Port. Meanwhile, I already had three world-class architects work-

Construction workers install the first steel beam for the
Freedom Tower on December 19, 2006.

ing away on their designs for the towers, their teams sequestered in the design studio I had set up on the tenth floor in Tower Seven, all of them committed to delivering blueprints by the date I had agreed to in the MDA, April 30, 2008. And not least, the economy in 2007, even going into 2008, was booming; it was arguably one of the best financial markets the country had ever experienced. It would be a propitious time to attract tenants as well as financing.

Seven years after the attack on Ground Zero, years filled with delays, acrimony, and spurious, orchestrated ceremonies, it finally looked like both the Port and Silverstein Properties were poised to take a giant step forward.

Except it didn't happen. The maddening problem was that while so many of the pieces were moving into place, a very necessary component was still missing. And the Port had done its disingenuous best to keep this a secret.

LIKE THE ORIGINAL WORLD Trade Center, the new complex would have an extensive underground interior highway system. It would allow all the site's mundane, unsightly, yet necessary day-to-day operations to occur below grade. Trucks transporting supplies, mail deliveries and pickups, passenger cars and tourist buses heading to the garages—all would use the ramps and facilities that would be part of this underground network. And the entire subterranean structure, of course, had to be built to withstand explosions and other security concerns. It would require a tremendous amount of concrete to harden these spaces. All doable, yet all time intensive.

In addition, the maze of below-grade train lines, subways and PATH, had to be engineered so that commuters could make their way to and from each of the towers on the site. And central to that complex train network was the transit hub, Calatrava's much delayed and increasingly expensive Oculus.

The ambitious design for the entire sixteen-acre site was an interconnected maze composed of many complex subterranean pieces.

Skidmore, Owings & Merrill rendering of the fortified glass, steel, and concrete base of One World Trade Center

Each element in this intricate infrastructure mosaic had to be completed, fitted neatly together in this underground netherworld, before the serious construction work could begin on the office towers and the completion of the 9/11 Memorial. And for a project of this size and scale to meet the loudly announced completion dates, it was essential that the work on the many belowground elements be delivered on schedule. If a single piece was missing, it could prevent all the other parts from moving forward in a timely way.

Every delay created big problems. Each postponement made nearly every aspect of the project much more expensive. It meant there were additional bills you needed to pay, more overtime expenses that were incurred. And it made planning a nightmare; try, for example, ordering the tons of steel necessary to erect a skyscraper, only you can't specify with any certainty when construction would begin. Or try attracting tenants, persuading them to make multimillion-dollar lease commitments, only you can't offer with any certainty a date—or even a year!—when they could move in. And, of course, I was still paying $10 million each month to the Port, $120 million a year, to lease sites that were sitting empty. A developer can't make money that way, funds going out, but nothing coming in.

So to avoid these sorts of problems, it had been specified in the revised 2006 agreement that the Port had to finish the components of the infrastructure on a precise timetable. They needed to deliver sites to me by the end of 2007 and 2008 so that I could fulfill my part of the agreement: I had to start construction pretty much simultaneously on three large office towers and have them "substantially completed" within four and a half years. If I failed to meet these deadlines, I would be subject to what the MDA called "cross-default." That meant I would lose the right to build any of the towers. It was one hell of a steep penalty, but I had accepted it because I had bought into the Port's rosy vision—five office towers, the Memorial, the transit hub—all up and running at the same time in the not very distant future. I believed they were committed to the written promises they were making. And I believed that they would be operating in tandem with me, quickly rolling up their sleeves and getting to work with me to make our shared vision a reality.

But as my team began preparing to start on the heavy foundation that would support the eastern part of the slurry wall—the "bathtub" that would hold back the encroaching Hudson River—I started getting reports that the Port was way behind schedule. In fact, the Port wasn't even ready to allow my crews complete access to the sites. And it wasn't just what my people were telling me. I saw it myself. I could look down from my office on the thirty-eighth floor of Tower Seven and see that not much progress had been made on the transit hub or toward the completion of the infrastructure. The Port wasn't even close to delivering the sites, as the MDA required, in "construction-ready" condition. Equally disturbing, it didn't seem to me that their crews were in much of a hurry to get this work done.

So I went to the Port, to the commissioners, and I began to ask questions. "We hear that there are problems," I said. "We hear that there are delays," I repeated anxiously.

"No, everything is fine," they assured me. "Just fine. We're right on schedule."

I wanted to believe them, but I knew that was not the case. In fact, I knew they were wrong. I mean, they were not even letting us into our sites to do the preliminary work because they still hadn't completed what they had been obligated to do.

And it was not just me to whom they were giving inaccurate reports. Tony Shorris, the executive director of the Port, told the press "by the end of 2012, it'll all be over but the shouting." Which was of course impossible. Then they met with David Paterson, who had requested a briefing on where things stood. And the Port's chairman told the new governor (the third since the project had started) that everything was going swimmingly, right on schedule.

Why had he said this? There was a very good chance he had been fooled, but I will always believe this sort of wild misrepresentation was in many ways a direct consequence of how things worked in the political world. You told people what they wanted to hear, what would make them happy. You didn't want to make waves, or say anything that would cause additional questions to be asked. And even if you finally told the truth, then it had to be kept secret. Buried deep from the press and the public. In April 2007, for example, the Lower Manhattan Construction Command Center, which was established by Pataki and Bloomberg, wrote a report stating that the underground facility for screening vehicles entering the site, as well as the PATH station and the memorial museum, were more than four years behind schedule and way over budget. The Port refused to make this assessment public. Apparently, they didn't want the public to know how their toll dollars were being promiscuously spent.

Private sector developers don't have that license. They can't just cover things up, pretend they are not happening. They have to deal in the real world of contracts and obligations. And they have to pay the bills with their own money.

And as for money, do you know how the Port, despite the constant revisions of the construction schedules, kept on budget? They simply ripped up one budget and replaced it with a new one. Higher costs didn't affect them; as they'd bragged, they could always raise tolls on bridges if cash was needed. They wasted billions; the site would wind up being $5 billion over budget. Yet it wasn't their money they were profligately throwing around, but the taxpayers'. This was the sort of undisciplined, free-spending mentality that I, a private developer, had to deal with in my partnership with a public agency. It sure made things difficult: I was locked in a constant battle with people who had neither the expertise to get the job done nor a sense of responsibility.

IT DIDN'T TAKE GOVERNOR Paterson long to discover that he wasn't being told the truth. He quickly learned that, despite what he had been informed initially, there was simply no way the site would be up and running by 2012. And now he was steaming.

His first step was to get rid of the Port's executive director; this was his appointment, while the governor of New Jersey selects the chairman. As the replacement, the new governor chose Chris Ward. Ward was a breath of fresh air, a very competent guy with an extensive background in public transportation and environmental work for both the city and the state, as well as a master's in theology studies from Harvard Divinity School.

Next, on June 11, 2008, Paterson gave Ward the job of writing a report that would be "a candid and transparent" assessment of how the rebuilding was progressing. In only about two weeks, Ward submitted a thirty-three-page report called *A Roadmap Forward* that read like an indictment: The 9/11 Memorial would not be open as promised for the tenth anniversary of the attack; the $2.5 billion PATH station and the Oculus were not only behind schedule and over budget, but the final design had not even been completed; and as for the twenty-six separate but interconnected projects that needed to be substantially completed before construction on the five towers could begin, "the schedule and cost for each of the public projects on the site face significant delays and cost overruns."

And who did the report blame for the completion dates that "were inaccurate from the day they were announced"? It diplomatically didn't provide names, but it clearly suggested, as *The New York Times* reported, that it had been Governor Pataki, City Hall, and the Port who had "wanted to convey the sense that this most politically, economically and emotionally significant of projects would be completed quickly."

Governor Paterson was similarly accusatory. At the press conference that followed the release of Ward's report, he said that the prior schedules and estimates had been so unrealistic, and so much essential information still remained unknown, that it would be impossible for the Port to set a completion date. But he was certain about one thing.

"Here's what we're not going to do," the governor emphatically stated. "We're not going to give any phony dates or timetables at this point and then follow it up with phony ribbon-cuttings and encouraging words and no follow-up."

And in case anyone had any doubts about whom Paterson had been throwing under the bus, the *Times* helpfully made it clear: "The reference to phony ribbon-cuttings and the like amounted to a rebuke of Mr. Pataki, who was in office in 2001 and viewed the rebuilding of Lower Manhattan as a central element of his 12-year legacy."

I felt a sense of relief that at last the *Times* was going after someone other than me. But when I was asked to comment on the report, I simply said with a measure of restraint, "As of today, my company's projects are fully designed and on schedule." I figured people would easily grasp the comparison I was drawing without my connecting the dots. In business, it was never wise to go out of your way to make enemies.

But I was also thinking that it didn't matter that I had so far fulfilled my portion of the MDA obligations. If the Port couldn't deliver construction-ready sites as they had agreed, then how would I meet my next set of deadlines? And if the cross-default provision was activated, then I would lose everything. While in the meantime, I couldn't finance or market 6.2 million square feet of space because there was no realistic date for when the Port would even let me into the sites to lay the foundations.

And there was another, even more infuriating complication. The Port was telling me that, in fact, they had finished preparing the sites for building. That they had fulfilled all their obligations. In October 2008, the Port issued "Certification of Final Site Completion" documents for both my Tower Four and Tower Two.

It was a total misrepresentation. Despite the Port's certifications of completion, the sites weren't ready for me to begin building. I was bleeding money. I needed to do something.

THERE WAS, I REALIZED, a way out of this mess. The 2006 MDA had not only established obligations for the Port and Silverstein Properties, but it had also created a mechanism for dealing with problems: arbitration. According to article 9 of the agreement, rather than our going through the costly and time-consuming process of filing lawsuits that could stretch on for years, a panel of three arbitrators, working on an abbreviated schedule, would "have plenary power to resolve any and all disputes." And I had made sure the MDA included a penalty with real teeth: the Port would have to pay Silverstein Properties $300,000 a day—the equivalent of the ground rent I was paying—for each day they were behind schedule. Earlier in the year, in fact, the Port had already paid a colossal $43.8 million in penalties to me for missed deadlines.

In November, incensed by the Port now claiming that they no longer needed to pay further penalties because they had delivered construction-ready sites a month earlier, I decided to initiate arbitration. What other recourse did I have? It was patently obvious that we couldn't begin work on the sites in the state they were in. There was, as a glaring example, a large wall still standing that prevented my crews from being able to complete the foundation. But demolishing the wall was not an option: it was a crucial support for the temporary box that surrounded and protected the tracks for the No. 1 train that ran through the site. Both the wall and the box would need to remain in place for months, perhaps even much longer, until the Port completed the belowground infrastructure. And that meant it would be a long time—a year or more possibly—before I could start building in earnest.

The three-person arbitration panel included a construction executive and a retired judge, and they heard testimony for three days. When it was over, the Port lost, hands down. The arbiters ruled that the Port's claims that the site work had been completed were false, that they must get back to work and finish their turnover obligations, and that they'd need to pay $300,000 a day in back penalties starting from when they'd stopped last month and continuing until the work was completed.

How did the Port respond to this thumping? They came to me and said with a shrug, "We knew we were going to lose." "Then why did you make me go through the time and expense of arbitration?" I challenged testily. "I spent a couple of million dollars in legal expenses" (fees that were not recoupable under the terms of the arbitration agreement). They explained that they just couldn't give in without putting up a fight. They explained that this is the way they operate. It would have been too humiliating just to give in.

And what about the money you now owe me? I asked. When added to the sum that had already been paid, it would eventually amount to more than $100 million in penalties. "Embarrassing," they said. Embarrassing? It was a lot more than merely embarrassing to the toll payers; this was money that could have been used to improve transportation, perhaps reduce tolls.

But once the arbitration panel's decision had been made, the Port came to me asking for a new favor. Not wanting to look stupid, they begged me to help them gloss it all over.

It had been, of course, a damn stupid waste of time, money, and effort for them, as well as for me. But I also knew I had to work with these people going forward. It was the future of the project that truly mattered, not my having a last laugh. In business, as I always point out, you can't afford to hold grudges. So I said, Sure, let's issue a joint press release.

"Both parties remain 100 percent committed," our mutual statement earnestly promised, "to delivering on our shared commitment to rebuild the World Trade Center. We are also pleased that in their ruling the arbitrators recognized the joint cooperation that has and will continue to exist between the Port Authority and Silverstein Properties in resolving the myriad of complex construction issues that have arisen and will inevitably arise as construction moves forward."

———————————————

FOURTEEN

I T WAS JANUARY 2009, and once again I was back at the beginning. I still didn't have any firm dates for when the Port would
deliver my construction-ready sites. I still couldn't make any
definitive plans about when I would be able to build, or, for that
matter, finance and lease my three towers. But even as I continued to
wait impatiently for the Port to live up to the commitments it had
made in the MDA, I had found a reason for at least a glimmer of
encouragement.

My small surge of confidence wasn't simply caused by the arbitration panel's ruling so completely in my favor. Or that since 2008 the
Port had paid Silverstein Properties more than $100 million in penal-
ties. My optimistic mood was generated by the fact that the Port, at
last, reached out, suggesting that we needed to cooperate if we were
going to get the massive task of rebuilding back on track. This was
precisely what I had been urging all along.

Chris Ward, the Port's new executive director, was the first to
reach out. Earlier he had issued the 2008 report entitled *A Roadmap
Forward* that wasn't so much a path to the future as a candid assess-

ment of how his agency had gotten lost, heading off at great expense down one wrong road after another. He now conceded directly to me that there was no way the Port could achieve the dates it had agreed to in the 2006 MDA.

"It's impossible," he began with an honorable candor. "It just won't happen." Instead, he said Silverstein Properties and the Port needed to make an entirely new deal. A deal where we would work together. "Because," he concluded, "here's what I've come to realize: we'll both need help if we're going to succeed."

It wasn't long after that heartening conversation that Tony Coscia, the chairman of the Port, also spoke to me. As if reading from the same conciliatory script, he said, "Larry, we can't do this without you. You can't do this without us. Therefore, we have to work out a whole new arrangement where we're working together to get this thing on track."

I listened to both men and I was immediately buoyed. What I heard was the Port, in effect, telling me, "Larry, you have the experience. You can do this quicker, cheaper, and better than we can. Let's figure this out together."

And, in fact, I already had a plan.

IT HAD BEEN A rocky time in America. The nation's economy throughout 2008 and now continuing on into 2009 was in free fall. It was a genuine recession, the capital markets in complete disarray. It couldn't have been a worse time to try to raise the $3.2 billion in financing I had estimated I would need to build at least two of the three towers I had promised for the site.

But with the help of my bankers at Goldman Sachs, I had come up with a plan in the summer of 2008 that would allow me, even in a recession, to get the financing I needed. It would be the most logical, expeditious, and cost-effective way—if the Port was truly serious about our working together.

Here's what I had proposed: I asked the Port to allow Silverstein

Properties to use their bulging balance sheet—money from tunnels, bridges, and airports—to backstop the loans we would get from the banks and financial institutions. We weren't, I made clear, asking for a handout of public money. Silverstein Properties would obtain the financing it needed from banks and commercial institutions, not the Port. In this unsteady economy, though, the only way we could borrow the billions we needed to build would be by using the Port's dependably enormous revenue flow to guarantee the loans.

Once we had the money, we would get the towers built. And as soon as they were going up, when people saw the gleaming skyscrapers designed by a trio of Pritzker Prize architects, technologically state-of-the-art buildings, energy efficient and environmentally sound, we would get them leased. Companies would move from Midtown because with all the tax abatements we had received, the rents would be a fraction of what they had been paying there. It would cost them less to have their offices in architecturally significant modern towers with walls of windows looking out on some of the best views on the planet. And if the buildings were occupied, then the Port could count on getting their monthly ground rent from Silverstein Properties. Which, after all, was the bottom line for the Port. It was why they had originally decided to lease me the properties back before 9/11.

But suppose, I went on, things didn't work out as rosily as I envisioned. Suppose I couldn't find tenants. Well, the towers would be up, and the Port would be able to lay claim to them at a cost substantially less than would have been required if the authority had built them on their own. The savings to the agency would be at least $2 billion. I would have already plowed the insurance proceeds as well as the money from the sale of $2.6 billion of Liberty Bonds into the towers—and that would be several billions dollars' worth of investment the Port would be inheriting. Even if I blew it, even if I couldn't get the buildings filled with tenants and had to default, then the Port would still receive a couple of fantastic real estate properties at a fraction of what they otherwise would have cost. And New York

City wouldn't be living with a large hole in the ground, a lingering scar from a terrorist attack years earlier.

It was a no-brainer, I told the Port officials. A win for you. A win for me. And a win for New York because it would create thousands of construction jobs and revitalize downtown, as well as create a fitting tribute to the memories of those who had lost their lives at Ground Zero.

So I shared all this with the agency's leadership. Think it over and then let's meet to discuss it, I suggested. After all, they had previously told me the Port now understood we had to work together. I figured they would be eager to sit down with me, but I never heard back from them.

I also explained my idea about the Port's backstopping my bank loans to Chris Ward. He listened attentively and then said he'd give us a counterproposal. That had been in September 2008. October, November—we had waited for his response. In December, we had finally received his document.

I had been expecting a precise outline for an entirely new schedule, one that would detail how the Port and Silverstein Properties could cooperate to finance and build all the towers in a timely and cost-efficient way. What I received was the furthest thing from that sort of practical, reasonable proposal. Instead, it was an irrelevant couple of pages. It basically told us to get lost. Four months, and this is what we got! And of course all this time we had been paying ground rent to the Port, $10 million a month, $120 million a year. Money that could have been used for construction. So much for all my newly found hopes. Once again I had been left exasperated, and fuming.

But I reined in my anger; in business, it was always counterproductive. And so my team began work on a reply. In four days, we had crafted a thoughtful critique of Ward's vacuous proposal. Our basic argument was that to build a single tower wouldn't be sufficient. A solitary building on Ground Zero would have a very difficult time attracting tenants. It would seem like a lone outpost on a wild frontier. It was necessary, we explained, to create a critical mass, to give people a sense that the entire site was moving forward toward completion. And we offered a compromise: Backstop just two of my three towers. Guarantee the $3.2 billion that it will cost to build these two skyscrapers. I would get around to the third tower later.

The initial response—silence. Finally, after several weeks, the Port contacted us. We need more information, they had said. More details.

We gave them everything they wanted. Again, we got it all to them in four days. And again they had us wait. And wait. We didn't get a response for another couple of months, months during which very little substantive construction proceeded on the site. But in April 2009, we at last received a new document from the Port.

It was worse than the one we had gotten in December: more recalcitrant and less specific. It certainly wasn't anything like *A Roadmap Forward* that Chris Ward had publicly touted.

Also troubling: in the spring of 2009, while I was growing increasingly infuriated by the lack of substantive response to my financing suggestions, the Port reverted to its old scheduling chicanery. It gave us a new series of dates for their completing the infrastructure, the 9/11 Memorial, and the Oculus that didn't make any practical sense. There was no way they were realistic. And if we didn't have a viable schedule, how could we begin to make construction plans, order steel, hire crews?

So we asked them for some information that would help us confirm their timetables. We're all working together, right? We all need to cooperate, as Ward had reiterated to me. There shouldn't be any reason not to share this information.

Only they wouldn't do it. They refused to give us the specific construction data that they had used to formulate this new timetable. And I was pretty sure I knew the reason for their reluctance: the new schedules were all wishful thinking, just another PR stunt. They didn't want to admit how they far behind they had fallen.

I wasn't the only one who was pointing out that the Port's latest timetable was completely out of touch with reality. The Lower

Manhattan Construction Command Center, an agency formed by
the city and state to monitor the progress at the Trade Center, had
issued a second report. It, too, found that "significant construction
delays" still remained.

I had had enough. Sure, the Port had already agreed to help
finance Tower Four, where it would be headquartered, occupying
600,000 square feet. (They chose not to take space in the Freedom
Tower after the agency's chairman had made the astoundingly unproductive statement that he wouldn't force the authority's employees
to work at a site that might be a terrorist target.) But that was not
sufficient. I couldn't raise financing in the midst of a recession, and
the Port was refusing to help backstop my loans on at least one more
tower. Besides, even if I had the money, I couldn't start building in
earnest because the Port had not fulfilled its obligation to deliver the
construction-ready sites. The entire situation was completely frustrating. And the Port's refusal to work with me, or even give me accurate
schedules, only made everything worse. I had come to the conclusion
that I had no choice but to take the Port to arbitration. Again.

But just as I was preparing to submit a letter of intent to file for
arbitration, the first step that was required under the 2006 MDA, I
received a phone call.

"LARRY," SAID MAYOR BLOOMBERG on that April afternoon in 2009,
"let's talk before you do something rash. Maybe there's a way we can
work things out."

"Mike," I responded, "this whole thing is really tragic. These guys
at the Port are spending buckets of money. Money that comes from
the citizens of New York. The Oculus is years behind schedule and
will wind up billions over budget. They've already paid me a hundred million dollars in penalties, but I still don't know when the sites
will be delivered. It's not incompetence. It's gross incompetence." I
said that I had no choice. "I'm going to head back to arbitration," I
announced with resolve.

All the time I had been venting, the mayor just listened quietly. Now he spoke up. "I can understand your frustration," the mayor agreed. "But," he urged, "give me some more time. Just hold off a bit. Let me see if I can solve this."

When the mayor of New York asked you to do something, it was only good manners to listen. But I was also thinking that when the mayor's an astute businessman like Mike Bloomberg, it made particularly good sense to pay attention. He had certainly demonstrated that he knew how to get things done in the real world; he had started a business from scratch that had made him one of the richest men on the planet. Maybe he could in fact help me work things out with the Port.

"Of course, Mike," I said. "I'd be grateful for anything you can do. I'll hold off on the arbitration letter."

So the mayor invited all the principals—Port officials, me and my key aides, the governors of New York and New Jersey, as well as Sheldon Silver, the powerful Speaker of the New York State Assembly—to a breakfast at Gracie Mansion in May 2009. It would be an off-the-record discussion, a give-and-take exchange where, the mayor promised, everyone could talk candidly. We could all air, he promised, our grievances. He wanted, he said, "to find a way to align incentives and keep progress moving at Ground Zero."

At the breakfast, we were seated around a long dining room table on a bright spring Thursday, the sun pouring through the windows, and the mayor started things off. His tone was calm, very reasonable. "This makes no sense," he began. "We've got to figure this out. There's got to be a way to get this done."

Nearly all the guests nodded their heads in mute agreement. *Maybe this will work*, I thought. *Maybe we just needed someone to bring us all together, to get us talking.*

"Silverstein," the mayor abruptly ordered. "You'll be cooperative."

"I'll cooperate, Your Honor," I agreed.

Then Bloomberg turned to Ward and Coscia, who were representing the Port. "You'll be cooperative, too," he said.

And they also agreed. "We'll cooperate, Your Honor."

He then moved on to a broad outline for a new plan. "Each of you will kick in some more money. And the state and the city will kick in some more money, too. We'll help you both get this done."

The mood in the room was, I felt, very conciliatory. Once again I started to believe that a solution could be found.

But when the discussion moved on to specifics, things rapidly began to fall apart. The Port angrily insisted that it could not allow me to use public money to build my towers. So I tried to explain that I wasn't asking for public funds. I just wanted to use the Port's balance sheet, the huge revenue stream that poured into the agency each month, to backstop the financing I would obtain on my own. It won't cost the Port or the public a penny, I carefully made clear.

Back and forth they went over the same old, well-trodden territory. No one was ready to yield.

The mayor's frustration became visible. He was clearly exasperated by the Port and me trading jabs only to get nowhere. "This is a result of September eleventh," he suddenly erupted forcefully. "We just have to figure out how to do this! Let's not find reasons not to do it. Let's find reasons to do it." Bloomberg was nearly pleading.

The mayor's words, however ardent, however earnest, could only do so much. The reality was that while the mayor had thrown himself directly into the mediation of our dispute, other than his bully pulpit, he didn't have any authority over the Port. It was the governors of New York and New Jersey who had the power to effect change at the authority. All Mike ultimately could do was make suggestions. The Port didn't have to listen.

Nevertheless, when he realized this discussion had become unproductive, the mayor made a new request. He asked that I continue to hold off on the arbitration for another three weeks.

But as June 11, the final day of this three-week period, approached, I told my people that we had clearly reached an irreconcilable impasse. There was only one way we would ever be able to get the Port to face up to the economic reality and work with us on the financing, I said. Prepare the papers requesting a new arbitration, I ordered.

But once again before the letter of intent could be filed, I received a phone call.

I WOULD LIKE TO meet with you, said Governor Paterson after I picked up the phone. Will you come up to lunch in Albany? Tomorrow?

It was a very last-minute request. I had a full day of appointments scheduled. But I wanted to believe that there still might be a way to find a solution.

Of course, Governor, I said. I'll be there at twelve thirty. And I added that I'd be bringing Janno Lieber. Janno was the head of my World Trade Center group. He had assembled the teams of architects and engineers. He also headed all the crucial negotiations with the Port to ensure that their work belowground would be fully completed, and then he helped finalize the street elevations for the towers. In fact, I can't imagine that Towers Three and Four would ever have been built without his staying on top of everything that had been involved in their construction. He was my point man, really, on everything connected to the project. A ceaseless and tireless guy, and incredibly smart: Harvard undergraduate and then NYU law school. Add to that, a real gentleman. Without him—forget it. (Janno now heads the city's Metropolitan Transportation Authority, which runs New York's vast system of subways, bus lines, railroads, and toll crossings.)

We arrived at the governor's mansion and he was there with Larry Schwartz, known as the governor's secretary, but really his chief of staff ("I'm like the three-hundred-pound offensive tackle blocking for the quarterback called Governor David Paterson," was how he once described his role). It was just the four of us for lunch, and the food was . . . well, you don't go to Albany for the food.

After only a few minutes of small talk, the governor took charge and got down to brass tacks. "We're going to have to come up with something different than what the mayor proposed. And you need to recognize something—without me, you don't have the state on your side. And it's the state that controls the Port Authority. This is where the power is. Right here."

"Governor, I understand," I replied. I mean, what else was I going to say?

"Okay," he went on equably, seeming satisfied that I'd acknowledged his position. "What are the problems?"

For the next half hour or so, I did my best to state my case. Then, having set the stage, I proceeded to explain to the governor what I was hoping for. Not a handout from the Port. Not a penny of public money. I simply wanted to use their balance sheet to backstop the loans I would need to build my towers. I told him the Goldman Sachs team who worked with us had already gone over it with the Port executives. They had carefully demonstrated that the break-even cost was so low that the agency's level of exposure would be very modest. And in the worst possible scenario, if I defaulted, the Port would take title to two world-class towers at a savings of billions of dollars. But that would never happen. It would be a win for everyone, including the citizens of the city and the state. It would jump-start the rebuilding of the entire site.

The governor had listened, barely interrupting. He now sat silently, as if weighing all I had said. "Here's what we're going to do," he announced finally.

He said he was going to let the Port know that he had met with me. And then he would call a press conference. He would lay out to the reporters everything I had said. He would explain to them why it made sense for the Port to backstop my loans.

That would shake the Port up, he said. That would force them to realize they have to work with you, he predicted. They wouldn't want to butt heads with the governor, he added with a confident laugh.

Janno and I were elated. We got into the car for the long ride down to New York City believing that the governor was sincere. In my ebullient mood, I could even forgive the lousy food.

About an hour later, as the car was winding down the Taconic Parkway, I got a call from my office. It seemed the governor had a press conference, all right. However, he had offered a completely different analysis than the one he had shared with us over lunch. He reversed himself 100 percent: he told the reporters that he agreed with the Port's refusal to backstop my loans.

I couldn't believe it! How could this have happened? To say one thing, and then do something completely different? It seemed incredible to me that the governor of New York would behave so mercurially, or that he would be so disingenuous.

Later, I learned what had occurred. It seemed that after we left the governor's office, Chris Ward walked in. And Ward, besides being executive director of the Port, also happened to be a friend of Larry Schwartz, the governor's influential chief of staff. The two friends, I suspected, had badgered Paterson until he changed his mind.

Paterson was the third governor (and not the last; there would be two more) with whom I had to work on the Trade Center. He was an accidental chief executive, a lieutenant governor who suddenly had to take over after Eliot Spitzer resigned in disgrace. Only he wasn't ready for the top job. He just didn't have the confidence to be the sort of leader the position required. He was too easily swayed by the last person to whom he talked. He didn't possess the intellectual capacity to look at a problem with objectivity and then think his way through to a carefully considered conclusion.

The next day, though, the governor and I talked. "What happened?" I asked. "What made you change your mind?"

"I thought about it," he said, "and I didn't want to look too prodeveloper."

"Governor, this is a total contradiction," I said evenly (yet I was raging inside). His sudden about-face still seemed inconceivable to me.

Paterson didn't try to apologize. Instead, he said, "Give me another week. I'll come up with some creative ideas. I have a couple of other thoughts already."

Did I believe him? I'm not sure. But he was the governor of New York State and I was a developer who did a great deal of business in his state. This was not the moment to hold a grudge. In fact, a half century in real estate had taught me that there was never such a moment. There was no room in business for personal animosities. You found the equanimity to put up with a lot of disconcerting things, and you never lost sight of what you were trying to accomplish. In the end, getting the project done, the deal made—those were the only things that mattered.

"Governor," I said, "if you're asking me to wait another week before I file for arbitration, I'm not going to say no to you."

"Just give me another week," he repeated. "I'll get back to you with some ideas in a couple of days."

A couple of days passed. Then a week. I still hadn't heard from the governor. In fact, he never got back to me. And I came to realize that Paterson was second-rate. He lacked substance. Sure, he meant well, but he couldn't make the tough decisions; he just wanted the entire impasse with the Port to go away as if by magic.

But it wouldn't. And in July 2009, I sent the Port a notice that I was beginning arbitration.

FIFTEEN

THE COUNTDOWN HAD BEGUN. Now that I had filed my letter of intent, the Port and I had ten days to try to reach an understanding or the arbitration panel would convene. That was the timetable that had been established in the 2006 MDA, and it had been designed to give us one final chance to iron things out on our own.

And there certainly was a compelling logic for both sides to want a last-minute attempt at reconciliation to succeed. The MDA had granted the three arbitrators "plenary power to resolve any and all Disputes." Further, ensuring that their verdicts would have real bite, they could "take such other actions ... necessary to enforce or implement a Decision." That meant the arbitrators' word would be law. We would pretty much have no choice but to abide by what three outsiders had to say regarding multibillion-dollar decisions. Neither the Port nor I looked forward to being in that helpless position—unless there truly was no other way.

Even after filing, I continued to hope that, in the days remaining before our fates would irrevocably be in the arbitrators' hands, the Port would come to its senses and try to work with me. And there had been one sign—admittedly a very small one, I concede—that suggested the Port was beginning to look at things at the Trade Center with a refreshing businesslike practicality. In March, the Port had decided to change the name of the Freedom Tower. Henceforth, it would be called simply "One World Trade Center." It was a rebranding designed to make the skyscraper more marketable to tenants— and less of a designated target for terrorists. The original overtly patriotic name had been another vestige of Pataki's self-aggrandizing legacy, and I was glad the Port had finally seen the wisdom of leaving at least a bit of that era, and its wrongheaded decisions, by the wayside. But was it a harbinger of greater compatibility?

Apparently not. The Port quickly made it clear that they had dug in their heels. As soon as the letter of intent was filed, Chris Ward lashed out with a statement claiming that I was "walking away from the negotiating table simply because the public has been unwilling to sacrifice critical transportation projects to subsidize private speculative office space."

Which was patently absurd. I wasn't, of course, asking the Port to subsidize any of my towers. I wasn't asking the Port to take a single penny from any of their transportation projects to invest in the construction of my buildings. I was simply asking them to backstop the loans I hoped to obtain. It would have cost them nothing.

Nevertheless, I didn't snipe back; I was still eager to move things forward and restraint seemed a more efficacious response. Anyway, I feared there would be plenty of time for that kind of vitriolic back-and-forth if we actually wound up facing each other in arbitration. For now, I wanted to keep the potential lines of communication open. I had my spokesman issue a carefully controlled reply. "Larry Silverstein is not walking away from the negotiating table," he told the press. And I hoped the Port was listening, too.

Mayor Bloomberg also tried to get things back on track before the deadline expired. "We are one step closer to a stalemate today," he said as the countdown began, "but a solution does exist and we will not stop working until we get there."

Yet even as he attempted to bring the two sides together, he didn't hesitate to point an accusatory finger. "Unfortunately, not everyone worked as hard as necessary to find a solution," he complained. And if there was any doubt about whom he was tacitly charging as responsible for this intransigence, the *New York Times* headline above the mayor's statement solved the mystery: "Port Authority Is Blamed for Trade Center Delays."

But despite our efforts, neither the mayor nor I could persuade the Port to consider a reconciliation with any seriousness. As time ran out, the dismayed mayor, in fact, released a statement assessing that we were farther apart than ever. And he was right. I'd been backed into a corner.

On August 4, 2009, following the procedures that had been outlined in the MDA, I presented the Port with a letter announcing that I was commencing hearings "seeking emergency interim relief." The fate of the World Trade Center would now be in the hands of the arbitrators.

THERE WERE THREE ARBITERS, the lawyer and the retired judge who had been members of the 2008 panel as well as a new addition, a construction industry executive. The hearings went on for fifteen demanding days, between October 12 and November 6, 2009. There were twenty-seven witnesses and hundreds of exhibits; if all the documents presented were stacked in a pile, I whimsically imagined it would be nearly as tall as any of the towers I hoped to build. It was a very exhausting, time-consuming, and expensive exercise. Yet what choice did I have but to proceed? The rebuilding of the site had so many complicated, interconnected parts with contingent construction obligations, and economic conditions had deteriorated so drastically in the two years since the MDA, that I felt only the arbitrators had the power to resolve the long-simmering disputes. Only they could force the Port to acknowledge the reality of this situation and the extent of their delays.

At the heart of my case was my fervent belief that the Port had completely bungled the rebuilding. It had not only failed to meet its scheduled obligations as detailed in the MDA, but also its announced timetable, I had come to believe, was "the product of the PA's misrepresentations and/or fraud." Those were strong words, but as proof I submitted the Master Development Schedule the Port had agreed to in 2008.

The Port's own timeline made the irrefutable case that the infrastructure elements were nowhere near the stages of completion that had been promised in writing. The Transportation Hub, the below-grade Vehicle Security Center, the underpinnings for the No. 1 subway line, and the reconstruction of Greenwich Street—none of these were close to being finished.

Equally disturbing, I had come to suspect that the Port had never even believed the promises they made. They simply announced completion dates that would look good in a press release. They had selected totally unrealistic dates to appease the politicians and the public, a calendar that would create the false sense that things were indeed moving along at Ground Zero. And I called a procession of witnesses, people who had worked at the Port in 2006 and 2007, who authoritatively confirmed the agency's cynical behavior.

The Port had deceived me. And this underhanded behavior was not simply a personal affront. It had costly consequences for my business. When the MDA had been signed, I had believed the Port. I had counted on the agency's doing what it had said it would do. I planned to go into the ground, to begin construction, in 2008. That meant I needed to buy steel, curtain wall, concrete, elevators in 2007—and this was at the height of a booming market. Prices were double what they would normally be. But what choice did I have if the agency was keeping to their timetable? Silverstein Properties couldn't be responsible for disrupting the schedule. Yet the Port let me do this knowing full well that they would never be able to deliver their infrastructure as they had promised. They stood by silently as I spent money for items I didn't need to purchase.

More fundamentally, though, I believed the Port had breached, as I put it in my pleading, "one of its most basic obligations" under our mutual agreement: its duty to cooperate. To get a massive project like this to the finish line, especially since so much of the below-grade infrastructure was interconnected, the Port and Silverstein Properties had to work together, each completely aware of the progress—or lack of progress—the other was making.

Only the authority had followed, as was stated in my filing, "a pattern of withholding critical information." And it was clear to me why their executives had acted so furtively: they had wanted to keep me "from learning that the project schedules could not and would not be met."

These were, I felt, serious violations of our agreement, and while I once again raised the issue of having the Port backstop my loans, I also argued for several additional remedies. For starters, I wanted damages that would offset the hundreds of millions in ground rent that I'd need to pay over the next decade; the site, I projected, would not be generating sufficient rental income from tenants until 2019 to justify the lease payments I was required to make to the Port. I also wanted the promised completion dates for my three towers to be moved back four years, until 2017. And since it was the Port's incompetence that prevented me from constructing my buildings on the schedule I'd originally anticipated, I wanted the cross-default provisions in the original MDA eliminated. Why should I be penalized and risk losing all my building sites when the Port was entirely to blame for the delay? I was the victim, I argued, not the guilty party.

Finally, yet not least, I wanted a formal declaration "that the PA is in material breach of the MDA." I didn't want this simply to point an admonitory finger or score points in what had become a very public battle. Rather, I thought that if the officials at the agency were embarrassed, chided for, as my filing put it, a "consistent pattern of delay and mismanagement," this would "serve as a wake-up call to the PA that it is accountable." It would be a well-publicized shaming that would convince the Port, I hoped, to get the project "back on track."

But I also knew this could prove to be wishful thinking on my part. Public opinion didn't seem to have much of an effect on officials in a well-insulated bureaucracy. Even if the agency were taken to task, it still might be unwilling, as I hoped, to "engage in good faith negotiations to reach a settlement." After all, their overwhelming defeat in the earlier arbitration had clearly done nothing to change their disingenuous behavior.

I feared that the agency's real strategy was to keep on delaying and delaying until all the insurance proceeds I had received would be spent on decades of ground rent and preconstruction costs. And when the insurance money to cover the costs of the lease was gone, they would really have me over a barrel. That would be when they would pressure me to accept their terms, or even once again attempt to force me out of the project entirely.

So I issued a warning in my filing. I wanted to put the Port on notice: if things didn't get moving once this arbitration had concluded, I would commence a new hearing "designed to obtain the full relief" to which I was entitled. And I let the Port know that the potential consequences of this third arbitration would be serious. I would be asking for damages of "at least $2.75 billion." I figured that would shake them to their senses; nearly $3 billion was a lot of money, even for them.

And how did the Port respond to my claims? The rebuttal arguments left me astonished. Chris Ward testified the completion dates in the MDA were simply "aspirational." The delays, he breezily went on, were "unavoidable." "Dates shmates," he even taunted at one point. Well then, I wanted to ask, was the contract I had signed requiring me to pay $120 million in rent each year to the Port also just an "aspirational" number? "Money shmuney," I nearly dared to sneer in response to his fatuous attempt at an explanation.

The Port also, no less incredibly, tried to put the blame for the delays on me. Their contention was that I couldn't attract tenants and financing because of the economic recession and the collapse of the commercial mortgage market. Yes, that was true. However, it need not have been that way. If the Port had adhered to its promised schedules, I wouldn't have missed the boom economy of 2006 and 2007; my towers, even if not completed, would have made sufficient progress to have attracted tenants ready and financially able to negotiate leases for future tenancy, since corporations required years to plan relocations. I felt that with this desperate argument, they were making my case for me.

But it didn't matter what I thought. The only opinions that mattered were those of the three arbitrators.

FINAL ARGUMENTS WERE HEARD on December 4, 2009. The MDA required that a decision be rendered within five days, but both sides agreed (for once!) that it would be an impossible deadline. There was no conceivable way the arbitrators could wade through the mountains of exhibits and the testimony from the long parade of witnesses and still make sense of all the complex and difficult issues in such a short time. Take as long as you need, we advised the panel. Nearly two months later, on January 26, 2010, they delivered a twenty-two page decision.

From my perspective, it was, for the most part, a disappointment.

Sure, the arbitrators pointed out that the Port "has not proceeded with construction of the infrastructure elements in accordance with the dates specified in the MDA." But they also posed a completely hypothetical, and to my mind irrelevant, question: given the financing difficulties, would I have been able to "timely proceed with and ultimately complete the construction of the Towers" even if the Port had delivered the sites as promised?

Then, rather than setting new dates for the completion of my towers, they said they wanted to wait until "more is known about the actual progress of the construction of the Towers and the infrastructure." Which seemed to me a cop-out. They were, in effect, deciding to delay making a decision because of the well-established pattern of delays. Only in the public sector would that kind of open-ended schedule be acceptable, or even envisioned. In the business world, where plans and budgets needed to be made and kept, this would be unthinkable.

The panel also refused to give me any relief from the three-quarters of a billion dollars in ground rents I would need to pay until 2019 when, I had estimated, my towers would as a result of all the years of delay be ready at last to welcome tenants. Instead, rather than grappling with the issue, they once again punted. The ruling was that I might in fact be entitled in the future to recover from the Port "excess" rent that had been paid. However, "at the present time there are too many uncertainties to permit a determination of either the length of the excess period . . . or of the proper amount of excess rent involved" that the Port might need to repay.

The arbitrators who had the responsibility to make a decision had again refused to do so. They preferred to wait until some unknown date in the future before dealing with this issue. But I had to run a business in the present, and pay bills in the present. Hundreds of millions of dollars were potentially involved. It was infuriating.

As for my request that the Port be cited for being in material breach of the MDA, that was denied outright. Although the panel conceded that "no doubt the first two years under the MDA do not represent fine, or even adequate, performance by the PA," they were "not persuaded . . . that the PA had acted in bad faith." Which I guess meant the arbitrators believed that the Port officials were simply not up to the job; their incompetence had not been deliberate but rather the best they could do. It had simply been my misfortune to be in business with people who did not have the skills to get the work done.

The panel also predicted that things would get better in the future. "Since Mr. Ward took over the directorship," they wrote, "much has changed, contracts have been let, and the infrastructure work is moving ahead." I could only hope this was true, since the Port had not provided me with the sort of specific information that would confirm this assessment. For all I knew, Chris Ward's efforts would also turn out to be simply "aspirational." The arbitrators, however, blithely felt confident that I could invest billions of dollars in the planning and construction of my three towers on little more than blind faith.

There was only a single ruling emphatically in my favor in the entire decision. The cross-default provision in the MDA was terminated; it was declared a relic of a "strangely naïve" time (which struck me as a "strangely naïve" analysis). I no longer had to erect my three towers simultaneously or risk losing my right to build any of them. At least something good had come out of the hearings. The specter of a cross-default had been a sword of Damocles hanging over my anxious head for years. I still had to fulfill all the other provisions of my original agreement, including paying rent, or I would risk default, but I no longer had to worry about trying to construct three huge buildings at the same time. It was a significant amendment to the previous agreement, and one that allowed me to breathe a sigh of relief.

But even my joy here was tempered by the panel's inconclusive concluding order. The arbitrators decided, in effect, to throw up their hands in perplexed dismay. Rather than issue a plan for resolving our disputes, they gave the Port and me "one final chance" to arrive at a solution on our own. We had forty-five days to come up with a new schedule for completing the three towers. And in that time we also had to figure out a compromise that would include backstop financing. If we didn't, the arbitrators threatened that they would impose a timetable of their own.

I was pretty much back to where I had started. I had spent all this time and money on the arbitration—and it had accomplished nothing of practical value except eliminating the threat of the onerous cross-default provision. All the financial penalties I had wanted imposed on the Port were shoved aside, to be decided at some future date, if at all. A new construction timetable had not been set; it, too, was put off for some unnamed future date. And as for the "wake-up call" I had wanted the panel to give the Port, well, the agency executives were more or less told they could keep on sleeping for at least another forty-five days.

Nevertheless, I tried to put on a bright public face. I released a statement that was carefully sanguine. "In this decision," I said, "the arbitrators directed us to work out a new plan to finish the WTC project quickly. That's a welcome development." And I concluded with a message I hoped would be well received: "I'm ready to work with the Port Authority 24/7 to hammer out a deal that assures the World Trade Center is fully rebuilt as quickly as possible."

Yet, in truth, I had little hope that our deadlock could be broken. I had fired my best shot when I convened the panel, and now that it had largely missed the target, I was dismayed. With as much objectivity as I could muster—and I concede that it was quite difficult to be unbiased when an investment worth several billion dollars was at stake—I still could not understand the logic of the arbitrators' decision. To me, it made little sense. Nevertheless, I was well aware of the bottom line: the Port had won, and I had pretty much lost. Even I, the perennial optimist, was beginning to lose hope. I was still determined that the Trade Center site be rebuilt. But how could anything happen during the next forty-five days, I wondered with a growing sense of desperation, that would get things back on track?

FORTY-FIVE DAYS. ANOTHER DEADLINE—and another chance to break through the impasse that had slowed construction at the site to nearly a standstill. I feared that unless I could come up with an inventive compromise on how many towers would be built and how they would be financed, then the arbitration panel would present its own plan. And I would be stuck with it.

That prospect left me unnerved. Judging by the mealymouthed decision the arbitrators had presented, I couldn't help worrying that their proposal would be so unrealistic that the site might never be up and running in my lifetime. I needed, I decided, to find a plan that would get the Port and me working together toward a reconciliation.

I was also being pressured by Governor Paterson to resolve my dispute with the Port. And while I had my problems with the governor, in this instance he and I for once had the same goal: we both wanted to see the sixteen acres at Ground Zero rebuilt, up and running, and busy with activity as soon as possible. We both agreed that there had been too many delays. Eight long years had passed since the attacks on September 11, 2001.

So in the aftermath of the panel's perplexing rulings and new deadline, when Paterson reached out and strongly urged me to try to reach a settlement with the Port, I listened.

"Throw some more of your money into the project," he blithely advised, as if I hadn't already committed significant funds. "Silverstein Properties should make a larger financial investment. That would placate the Port," he predicted. He also wanted me to scale back my ambitions. He hoped I would reconsider my plan to build three towers at the site.

I gave it all some thought, and in early February, about two weeks after the panel's decision, I proposed a new strategy for moving forward. I would put aside (at least for the foreseeable future) my plans for Tower Two. It had been designed by Lord Norman Foster, a 3-million-square-foot skyscraper that would be taller than the Empire State Building. It was a sleek, beautiful building, but in order to jump-start the entire project, I would put my plans for this 1,270-foot tower on hold. Besides, it was too expensive an undertaking at that moment, in that marketplace.

I would also invest, as suggested, further money in the remaining sites along Greenwich Street. I would commit $250 million—$175 million more than my previous offer—to the construction, and add an additional $560 million from the insurance and Liberty Bonds proceeds.

These were major concessions—and major sacrifices. But I was determined to show my commitment. And I wanted to believe the Port would behave this way, too.

I was mistaken. A *Daily News* article accurately summed up the belligerent response. "The Port Authority pooh-poohed the Silverstein proposal as insignificant," the paper reported. And it quoted "one source familiar with the agency's thinking," who dismissed my suggestion as "a non-starter." The *News* explained, "The Port Authority bureaucrats had demanded a lot more 'skin in the game' from the builder."

More skin in the game? I had already spent about $2 billion on ground rents, building designs, and preliminary construction. Despite my contractual right dating back to my original purchase of the leasehold, I had given up One World Trade Center, as well as hundreds of millions in insurance proceeds, to the Port. I had offered to put Tower Two on hold. And I had just agreed to invest an additional $810 million. That sure seemed like a lot of "skin in the game" to me.

What more could I do? I despaired. What would have to happen before the Port would come to its senses? What sort of catalyst would be required to reignite our negotiations?

I could never have guessed the answer that would soon come my way.

UNKNOWN TO ME, THE event that would change everything had already begun to take shape earlier that fall. Scott Pelley, the veteran CBS newsman, had been walking through the twisting maze of corridors that ran through the *60 Minutes* offices on West Fifty-Seventh Street and happened to cross paths with Shawn Efran, one of the news show's producers.

"Hey," Pelley offhandedly asked, according to the account he would write, "whatever happened to Ground Zero?" The question popped up in his mind because the two journalists had raced downtown together just minutes after the first 767 had smashed into the

Trade Center. At the end of that long, horrific day, Pelley found solace, he would say, in what had seemed a reassuring certainty: "Ground Zero would be rebuilt into a soaring statement of the American spirit."

And now, running by chance into the producer who had accompanied him downtown on 9/11, not only were all those memories recalled, but he also began to wonder just what had been done at the site over the past eight years. "A lot should have been finished downtown," he thought to himself. "Why wasn't I hearing anything about it?" What had become of that "soaring statement of the American spirit" that was going to be built?

In the aftermath of this accidental encounter, as Pelley's reporter's mind filled with nagging questions, the idea for a *60 Minutes* segment began to take shape.

WHEN THE REQUEST CAME from *60 Minutes* to meet with Scott Pelley and film an interview, my executives suggested I shouldn't even return the call. Don't get involved, they advised. *60 Minutes* only does hatchet jobs, they warned. Do you really want to go head-to-head with Scott Pelley? You think an investigative reporter is going to take the side of a businessman? A rich developer makes a perfect target, an easy fall guy. Remember, they pointedly reminded me, how the *Times* editorial board had consistently twisted the facts. You tried to educate them, gave them the full charm offensive during a lunch at the paper, and what had it gotten you? A lot of smiles, and the next day the editorial writers went after you with the same old misguided venom. You think Scott Pelley will be any different? Why look for trouble?

I could see their point. I certainly had taken a beating from the press. And the prospect of being unfairly pilloried on national television was daunting. Did I really want to go looking for trouble? It made sense, I could see, to ignore the request.

But I also knew that regardless of what I decided, *60 Minutes* was going to do a story. I had heard Chris Ward had agreed to be interviewed. Did I want the Port to tell their version of the events without my having the opportunity to respond?

Yet, more importantly, I knew I had a good story to tell. The arrogant mismanagement, the wasteful spending of billions of taxpayer dollars, the years of infuriating delays—I wanted to let the American people know just what had happened over the past eight years at the World Trade Center site. I wanted to give them the unvarnished facts.

"Let's do it," I told my people. "Call them back and set a date."

THE GUARDS WOULDN'T LET us onto the site. That was how my tour of Ground Zero with an astonished Pelley and his camera crew began. It didn't matter that I held the lease on the property and had been paying between $6 and $10 million a month for nearly a decade—about three-quarters of a billion dollars in total—to the Port as rent. The agency security force refused at first to give me access to my own property.

But the 60 *Minutes* crew had their cameras rolling. And on Sunday night, February 21, 2010, the estimated 11 million households watching the news show saw the incredible footage of my being turned away from my own building sites. At the same time, they heard Pelley's perplexed voiceover: "The hold-up at the gate is a symptom of how much the relationship has soured between Silverstein and the government agency that is supposed to be his partner."

These words set the tone for the newsman's no-holds-barred indictment that followed.

As the cameras panned laboriously over the vast, muddy site, Pelley noted that "this is Ground Zero today, much of it still a pit where there are supposed to be five skyscrapers, a memorial, a museum, a theater, and a transit hub." Then he turned to me and solemnly asked, "So when you look out on where this project is after eight years, how would you describe this?"

I didn't hesitate. I didn't reflect. I spoke straight from the heart. "I describe this as a national disgrace," I responded in an anguished outburst that would be quoted prominently in newspapers all around the country the next day. "I am the most frustrated person in the world."

I didn't stop there. "It's hard to contemplate the amount of time that's gone by here, the tragic waste of time and what could have been, instead of what is today."

Nor did I hesitate to put the blame directly on the Port. "There's no accountability," I complained. "And when they see they're missing a date, they publish a new date and say 'we're on schedule.'"

I went on, though, to make it clear that I hoped the Port and I could come up with a plan before the deadline the following month, when the arbitration panel threatened to create its own construction schedule. "You just want to work together, to bring this to fruition. To bring this to a successful conclusion," I said, trying to sound hopeful.

But Pelley saw through my thin expectations. "But the fact of the matter is you and the Port aren't talking to each other anymore," he countered.

"Well, we're talking but not in the fashion that one would expect or hope for," I replied with candor.

"Talking through your lawyers?"

"Unfortunately," I agreed.

With a bristling frustration that, I felt, mirrored my own, Pelley forcefully went on, "Instead of a construction site, Ground Zero became a stage for elaborate but meaningless ground breakings and ribbon cuttings." The blame, the reporter made clear, lay with the politicians and the bureaucrats. "Since 9/11 there have been three governors of New York, four executive directors of the Port Authority, and no one to see the project through."

He bluntly concluded: "On the ten-year anniversary of September eleventh, seven billion dollars will have been spent, but not one project will be finished. Most of the buildings in the master plan are in doubt and at best, a decade after the attack, Ground Zero will look [like an empty hole]."

60 Minutes anchor Scott Pelley tours the World Trade Center
site with Larry Silverstein in 2009.

It was precisely what I had been complaining about for years to
anyone who would listen. But now Scott Pelley was saying it on a
Sunday evening to an attentive audience that numbered in the tens
of millions. People would, I wanted to believe, now pay attention.

IT WAS QUICKLY DEMONSTRATED that at least one person had been
listening. Mayor Bloomberg had seen the *60 Minutes* segment and it
left him enraged.

On his weekly radio show, he erupted. "It's time to stop this crazi-
ness ... we just have to move forward.

"If we don't, you are going to see me out there beating the drums
every day. I'm not going to leave this world with a hole in the ground
ten years from now," he pledged.

The mayor had been well aware of what had been happening for years. He had started out being critical of me, but in time he had begun to understand my frustration with the Port. He had even convened a summit meeting at Gracie Mansion to try to move things forward, but it had failed to get things back on track. After seeing the 60 *Minutes* segment, however, his anger and disappointment had coalesced into a renewed determination. Was he a politician pragmatically responding to what was now a very public scandal? Was he a New Yorker whose irritation and exasperation had reached its limit? I don't know. All I know for sure was that his vow helped change everything.

SIXTEEN

YET AT FIRST EVEN Mike Bloomberg had trouble getting people to listen. As the weeks ticked away before the arbiters would set their own timetable for construction, the mayor kept beating on his drum, as he had promised. But he couldn't help feeling that no one was paying attention to the noise.

"Larry," he complained, "I don't know how to help." Sounding weary, full of a nearly palpable frustration, he said that while the Trade Center was located in Manhattan, he had neither the official authority nor the political clout to control the decision-making process. The Port, he went on, was a bi-state agency directed by the governors of New York and New Jersey, not the mayor of New York City. "I don't have the hot buttons," he said with a sigh.

"Yes, you do," I corrected confidently.

Frowning, clearly perplexed, the mayor fixed me with a dismissive stare. "Like what?" he challenged finally.

"The ARC," I said. Access to the Region's Core, or ARC as it was universally known, was a proposed $8.7 billion construction project the Port was hoping would increase commuter-rail passenger service between New Jersey and Manhattan. And a crucial component of this many-faceted infrastructure package was a tunnel beneath the Hudson River, a natural border that separated the city from New Jersey. "And you," I pointed out to the mayor, "control the New York end of the tunnel."

If he was to let the Port know that his support for the ARC tunnel was wavering, I suggested mischievously, but at the same time also make it clear that he believed the three World Trade Center towers should be built simultaneously, there would be no need to connect the dots for the politically astute agency executives. They would understand that in the tit-for-tat world of intragovernmental cooperation, a deal was being proposed: the mayor would overcome his reluctance to the tunnel if they would abandon their vehement objections to the financial backstopping that would enable the towers to be built simultaneously.

"Let me think about that," the mayor said cryptically. He had, I knew, already enmeshed himself in a great deal of controversy by pushing through an extension to the city's previous two-term limit rule. Perhaps, I feared, he would try to avoid additional squabbles during this third term.

But a week later, he made his thoughts very clear. On his weekly radio show, during a conversation with John Gambling, the WOR interviewer, the mayor seemingly ad-libbed a very carefully considered statement.

He began by acknowledging his frustration with the absurdly slow rebuilding of the Trade Center. "It's been going on for nine years," he railed. "We've got to get this done!" But, he continued as his voice rose up in dismay, the Port Authority was not as concerned about the lack of progress as he was.

His raw anger, however, was just a prelude. In the next moment, Bloomberg proceeded to detonate a well-planned verbal threat. Until the port was ready to help move this project forward, he announced, "I don't have any interest in proceeding with the ARC tunnel going to New Jersey. It's just going to have to wait."

Not more than four hours later my phone started ringing. I received calls from Port executives who hadn't spoken to me in years. Even from some who had never previously deigned to speak with me at all. And each and every one of them delivered the same resigned message: Let's sit down and hammer out a plan.

AFTER A SIXTEEN-MONTH STALEMATE, on March 25, 2010, the Port and Silverstein Properties reached a tentative agreement. And up to the very end, the politicians, shaken by the public airing of the long-running fiasco on 60 Minutes, were applying pressure on me to close the deal. Governor Paterson had telephoned me the night before I signed off to urge, "Stop bargaining. Accept the deal." The first thing the next morning, Mayor Bloomberg had called with a similar plea. And coaxed by their last-minute prodding, I accepted the Port's new development plan.

The terms, in broad strokes, were these:

I would get to build Tower Four, Fumihiko Maki's elegantly minimalist seventy-two-story building, which would contain almost as much office space as the Empire State Building. And with about 60 percent of the tower preleased to the Port and the city, the two entities renting a combined 1.2 million square feet, I would have tenants with first-class credit. I could count on both of these public entities to pay their rent. Satisfied by this reliable annual cash flow, the Port agreed to guarantee the issuance of about $1.3 billion tax-free Liberty Bonds, which would be sold to the public to finance construction—which was precisely the sort of backstop I had been asking for all along. I would now have the ability to finance Tower Four.

In fact, I would be going down a familiar, and proven, development path. I would be proceeding as I did immediately after 9/11, eight long years earlier, with Tower Seven. However, instead of using the insurance money I had received, I would be relying on the Port-guaranteed Liberty Bonds for construction financing. And as with

Tower Seven, where Silverstein Properties was the only initial tenant, I would be building largely on spec; I didn't have any of the space rented in advance. Yet that gamble had worked out. I had weathered the financial storm, built my skyscraper, and then, as I had predicted, the commercial market had improved, and tenants had flocked to the space. The building was already turning a profit. I was confident I would have similar success with this new one, too. Once Tower Four started going up, I felt sure I would be able to attract additional tenants. Then I would have no trouble getting permanent financing from the commercial lenders, and the Port could go off on its merry way, no longer needing to worry about the guarantees the agency had made to bond holders. The entire process would wind up not costing the agency a dime. It was a deal that made all the sense in the world for both me and the Port: I would get my building, and the agency would get its very significant flow of ground rent each year. If everything went according to the plan, I could have the building done in three years, by 2013.

The deal for Tower Three, the 1,079-foot-tall Richard Rogers building that would be situated between my two other skyscrapers on Church Street, was more complicated—and more problematic. I would build and pay for the seven-story podium base that would house the retail stores and the PATH transit infrastructure the Port needed to operate the terminal, but there were conditions. Before I could receive the additional money I needed to construct the remainder of the tower, I would have to raise $300 million in cash and find a tenant for at least 400,000 square feet. If I fulfilled both these requirements, the city, the state, and the Port would provide up to $600 million in assistance—again, the sort of financial backstop I had been asking for.

I accepted these conditions because, as always, I was optimistic about the future of the economy and the city. I estimated that it would take me not more than two years to build the base, and by then I would be able to find both the tenants and the capital I needed. After all, they were only requiring I lease 400,000 square feet in a tower with 2.5 million square feet of rental space. That was just a small fraction; I could easily get that done. And then I could keep on building upward.

As for Tower Two, the tallest and northernmost of my three, the Norman Foster tower with its slanting roof angled toward the 9/11 Memorial, I would build it up to street level under a financing plan that still needed to be worked out between the Port and me. But to build higher, I would need to wait until the credit markets improved and I could find a tenant that would allow me to obtain the money I needed for construction; I wouldn't have the luxury of Liberty Bonds backed by the Port to provide financing. Nevertheless, I figured that as long as I was proceeding on the other two towers, people would get the sense that things were moving along at the site and I would find a tenant for Tower Two. It was a state-of-the-art building in a great location; companies would want to move here. In the meantime, while I waited for the economy to gather steam, at least the site would no longer be an empty hole in the ground. I would have a seven-story base that was largely paid for by the Port. And while the buildings were rising, I knew the neighborhood that had been battered by the events of 9/11 would start to find a renewed life, too. With the promise of people returning to the Trade Center site, a vibrant community would begin to spread around it.

All in all, it was not only a deal I could live with, but it was also the sort of backstopping arrangement I'd wanted all along. An analysis in the *New York Observer* summed up the results quite accurately. "The deal comes as a victory for Mr. Silverstein," wrote Eliot Brown, "who drew a line at two towers, demanding the Port Authority owed him the responsibility to make it happen on account of the agency's delays. Mr. Silverstein was in a position of great leverage, as without a deal, infrastructure on the rest of the site would have been rendered non-functional without hundreds of millions of dollars in changes, and, if he defaulted, the Port Authority would have stopped receiving situation rent payments, currently around $80 million a year."

Yet, despite my "great leverage," despite the commonsense logic reinforcing what I had wanted, it had taken nearly a year and a half of acrimonious haggling with the Port, as well as an expensive and time-consuming arbitration hearing and a biting critique on the nation's most watched television news show, to get the agency to see things in a reasonable, objective way. I had to create a crisis, institute an arbitration—the second hearing, in sad fact—to get a legitimate, thoughtful response from the agency. And this might have very well been unproductive if not for the *60 Minutes* segment. These were, I had come to learn, the only sort of catalysts that would make government bureaucrats begin to function. Without a crisis situation, they felt no rush. They would just keep on dawdling and dawdling, making outrageous and ill-considered demands. And all the while millions of dollars of public money would continue to be promiscuously wasted, and the sixteen-acre site would remain a muddy hole.

But now wasn't the time to complain, I also realized. What was done was done. "This is great news for New York," I rejoiced in a statement I released to the press, and I meant it.

OR WAS MY CELEBRATORY mood premature? Over the years in my dealings with the Port, I had learned that an agreement was not a finalized deal. And what we had negotiated—the broad outlines of a settlement of our disputes—still had to be put into the precise lawyerly language of a contract that officially amended the 2006 Master Development Agreement. The Port's board gave Chris Ward, the agency's executive director, 120 days to work out this binding document.

It didn't take long for some of the Port's advocates to start having second thoughts. They were, apparently, not ready to surrender without putting up more of a fight. In a private session, the agency's executives, according to the report in *The New York Times*, "insisted that Mr. Ward get strong language limiting Mr. Silverstein's development fees while public money was at risk in the second tower."

In other words, even though they had caved on backstopping, the agency officials still were hoping to limit their previous financial commitments to me. Maybe, I feared, they were even determined to upend the entire deal.

And my fears were prescient. For no sooner had the details of the proposed resolution been announced than one of the agency's commissioners began a series of attacks. Henry Silverman, an abrasive businessman appointed by New York's Governor Paterson, grumbled to his fellow commissioners and to reporters that the terms—providing $600 million to help construct Tower Three, guaranteeing the funding for $1.3 billion of Liberty Bonds for Tower Four—were too exorbitant. The Port couldn't afford it.

Next the New Jersey governor, Chris Christie, chimed in, seeming to side with him. "While I am heartened and supportive of the broad outline that is leading to the conclusion of negotiations between the parties," he said in a cautionary statement to the press, "I remain steadfast that any final agreement protect the taxpayers, commuters and toll payers of New Jersey and New York."

This was not the ringing endorsement of the deal that I had hoped for. I knew that at the August board meeting it might take only one loud, dissenting voice to unravel all that we had finally accomplished.

Nevertheless, I was encouraged by the fact that Mike Bloomberg had promised me that he would make sure the exhaustingly negotiated deal would be approved. And he seemed determined to keep his word. At the press conference announcing the preliminary agreement, he very publicly continued to put pressure on Chris Ward.

Standing next to Ward, with all the participants listening, the mayor pointedly hectored, "Chris, you're not going to screw this up, right? You're going to get this done in one hundred and twenty days, right? Because there's a tunnel to New Jersey that's not going to get done if you don't. You understand that, right?"

Before a flustered Ward could respond, the dogged mayor focused in on another Port executive, a man directly involved in the ARC tunnel development. "You understand, we need to get this done," the mayor reiterated, barely bothering to conceal the threat.

Then, not waiting for the blindsided bureaucrat to respond, Bloomberg was quickly back in Ward's face, fixing him with a hard, relentless stare. "This is going to get done, right, Chris?" the mayor challenged again. "We're not going to screw this up?"

"Yes," a cowed Ward agreed. "Yes, we'll get it done."

But despite the executive director's assurance to the mayor, I couldn't help but feel that it would be a rough 120 days before the Port board meeting to vote on the issue. Four months was a long time. Long enough to undo a deal that still needed to be finalized. Long enough, I worried, for anything to happen.

AS THE DAYS COUNTED down to the Port meeting scheduled for late August 2010, I tried to remain confident that the Board of Commissioners would approve the negotiated agreement. Sure, for nearly two acerbic years, the commissioners and Silverstein Properties had gone at each other tooth and nail. There had even been periods when we weren't talking. But in the end, after all the tempestuous haggling, I had given up a lot, and so had the agency. And that, as I always say, was how deals got done.

Yet I also knew that I had been burned in the past; it was difficult to predict what the agency would decide. Part of that problem, I had learned after all the frustrating years, was that I never knew with whom I would be ultimately dealing. Or who had the final say.

At the time, there were ten commissioners (including the chairman and vice chairman) and nobody spoke, it seemed, for anyone else. Yet at the same confounding time, the commissioners were often directly influenced by the governors who had appointed them—and the heads of New York and New Jersey were rarely in agreement. In fact, they were more often than not rivals, ambitious competitors vying against each other for federal funds and headlines. And making the decision-making process even more unpredictable, there was always the possibility that the governor who had power today would be gone without much warning tomorrow; during the past nine years, as I struggled to get things moving at the World Trade Center site, two governors—New York's Eliot Spitzer and New Jersey's Jim McGreevey—had abruptly resigned because of their involvement in sex scandals. Suddenly, people who had little (or even no) prior knowledge about the complex factors involved in the rebuilding were empowered to make very consequential decisions. It wasn't at all like dealing with the private sector, where there was continuity, where judgments were generally based upon reasonable, bottom-line determinations rather than capricious politics. At Silverstein Properties there was only one person who ultimately made all the decisions—me. I never hesitated to make my opinions very clear. And I certainly didn't waver.

Yet while I continued to worry about how the board would vote, I did find reasons to have a bit of confidence. For one thing, Mayor Bloomberg continued to reiterate his support. "This is my legacy," he would tell me. "It's my final term. We've got to get this done." Equally reassuring, he had backed up his words by putting Bob Lieber, the deputy mayor for economic development, in charge of coordinating the city's activities at the site. Usually I found that, as many in the private sector instinctively felt, people who worked in government were just not up to the job; if they were any good, they would have gravitated to private industry, where they would get paid commensurate with their expertise. Lieber, however, was an exception: a municipal official who was both smart and dedicated. He put a lot of hours into the rebuilding and he was constantly pushing for things to go forward (unfortunately, though, he'd leave city government just two months before the board's vote).

And, another encouraging sign, Chris Ward had actually called me. This was a surprise in itself; during the second arbitration hearing things had gotten so acrimonious that I doubted we would ever talk again. It still stuck in my craw: Did he really believe that delivery dates in a signed contract were simply "aspirational"? Only now he was on the phone, telling me he was committed to getting the agreement approved and the towers built. Hearing those reassuring words, I was willing to let the past remain past. I told him, "If you're working to get things going, I'll be right by your side. We'll get it done together." And I meant every world of that pledge.

In the meantime, though, the future of the site hung on how the board would vote: without the backstopped financing, the towers would never be built.

Finally, after months of waiting, on August 26, 2010, the commissioners met and voted. The March agreement was approved. And with that, a potential $1.6 billion in Liberty Bond proceeds were now close to being available. I would have the money I needed to build two towers. At last, I rejoiced, things would move forward. I couldn't wait.

BY OCTOBER, TOWER FOUR, Maki's elegant, sleek skyscraper, had been built up to the tenth floor of an eventual seventy-two. But to keep the construction going, I was counting on the proceeds from the sale of over a billion dollars' worth of tax-exempt Liberty Bonds. Yet I was confident that selling the bonds wouldn't be a problem: Goldman Sachs and JPMorgan Chase, two first-class financial institutions, were handling the issuance, and the bonds themselves had received high credit ratings thanks to the Port's cash flow and its reserves that, after all my persistence, now guaranteed the debt.

But then without warning, the municipal bond market fell into turmoil. Worse, when rates settled, Fidelity, the giant Boston-based financial services concern, announced it had a problem with the impending bond sale. They were a major holder and seller of the Port's previous general obligation bonds and they objected to the agency's guarantee of the Liberty Bonds. If the Port hit rocky times and had to default, Fidelity asserted that the agency would still need

4 World Trade Center, designed by Fumihiko Maki, rises in 2012.

to pay its general obligation bond holders first, not the owners of the Liberty Bonds.

The Port was deeply embarrassed, but why hadn't they thought this through before? And as spring and summer had passed without the bonds being marketed, the insurance proceeds I had been using to construct Tower Four were rapidly dwindling; it had been costing me $50 million each month. I needed the bond proceeds as soon as possible, or else I would have to pause construction on Tower Four, and that would cast a pall over the entire project. I would need to lay off three thousand workers, and the cost to restart building after a long hiatus might very well run into the hundreds of millions of dollars.

It was not until the first week of November, two years after the Port had first agreed to back the bonds, that the Port's board voted to approve a new version of bonds. They went to market immediately, and the 4 World Trade bond offering was a tremendous success. It was oversubscribed eight to one. At last I had the money I needed to get Tower Four completed.

BUT I HAD NOT forgotten what I had had to go through to reach this victory. I could not help thinking about all I had to endure, the nearly constant pushing and shoving that had been required to move a complacent and wasteful Port Authority toward a responsible timetable. The time-consuming and costly arbitrations, the venting of my frustrations on *60 Minutes,* my wrangling with the mercurial politicians and the Port executives—all of it had been part of a long-running nightmare. Trying to work in partnership with the public sector had proven very demanding. It was an ineffective and all too often unreasonable way of getting things done. Politicians were too capricious; they didn't think like businesspeople.

Thankfully, it was not the way I usually did business. And the effectiveness of my development strategy, the way I had conducted business for half a century or so, had once more been proven by the successful sale of the Liberty Bonds. My Tower Seven, the skyscraper I had put up by myself, without the Port's intervention, five years after 9/11, was now an undisputed commercial success. I had built it entirely on spec, without a single tenant other than Silverstein Properties. Critics had argued that no one would return to the World Trade Center neighborhood. Even the mayor had carped that I would never get the rents I was charging. But now Tower Seven was fully occupied—and the tenants were paying the rents I had set. My unwavering confidence had reaped tangible results. In fact, I was in the midst of a refinancing that would allow me to pay off the original construction bonds. Tower Seven, as I had predicted, had become a genuine moneymaker.

It was a successful strategy I knew I could repeat if I was given the opportunity. And now that I had the money I needed to build Tower Four, I felt that nothing would stop me. I would build up and up until all three of my towers were aligned along Church Street and rising high into the sky.

PART III

BUILDING UP AND UP

SEVENTEEN

B UT WHILE THE REBUILDING of the complex was dogged by delay after exasperating delay, there was one component of the original master plan that was ready two years earlier than expected. Mayor Bloomberg had been determined that the 9/11 Memorial should be open in time for the tenth anniversary. He did not want to be standing in a scrum of politicians at another carefully rolled out ceremony only to be once again facing a gaping vacant hole. He did not want the television cameras focusing on where the Twin Towers had stood—the sacred heart of Ground Zero—and broadcasting the grim image of a muddy pit. Nor did he want the nation to be reminded how little the politicians and the bureaucrats had accomplished over the past decade. And through the power of his political capital as well as the capital in his own pocketbook, he succeeded in getting a substantial part of the memorial up and running in time.

Yet like everything that took place on the site, the building of the memorial had been a battle: another acrimonious tale of disputes, delays, mismanagement, and wasteful spending. And the mayor's triumph was only achieved after inventive and expensive changes to the construction schedules for the rest of the complex.

THE IDEA OF ESTABLISHING a permanent memorial where the Twin Towers had stood had been integral to Daniel Libeskind's original master plan for the sixteen acres. And in 2003, just two raw years after the terrorist attacks, the Lower Manhattan Development Corporation, the municipal group that in the last days of the Giuliani administration had been established to oversee the revitalization of downtown, launched an international competition to select a design. There were 5,201 submissions from sixty-three countries, and thirteen distinguished jurors sorted through them, searching for a memorial that would honor the victims, satisfy the poignant concerns of the families who had lost loved ones, as well as provide an oasis in the midst of a thriving commercial complex for healing and reflection.

The winner was Michael Arad, a young, unknown Israeli American architect who from the rooftop of his East Village apartment building had witnessed United Airlines Flight 175, the second of the hijacked jets, roar into the South Tower. His proposal, Reflecting Absence, was a somber, minimalist evocation of grief, yet it possessed a quiet power that could, the judges felt, dominate the entire complex. And its size, too, would be significant. It would be spread across eight acres, occupying about half the site.

Set on the footprints of the original two towers would be twin square reflecting pools, each below ground and nearly an acre in size. The pools were to be fed by waterfalls on all four sides (the largest man-made waterfalls in North America, as things worked out), each cascading down thirty feet into a dark, square basin before the water poured into a smaller central square and disappeared. Arad said the pools represented "absence made visible," the water perpetually flowing into a space that could never be filled. It was a very affecting metaphor for a previously unimaginable national tragedy.

I had looked at the design, and although it was still only a sketch I

The World Trade Center Memorial Park designed by Michael Arad
and Peter Walker features waterfalls where the Twin Towers stood.
The 9/11 victims' names are featured around each pool.

was struck by its melancholic power and dignity. I felt it would offer
not only a striking and appropriately solemn tribute to the 2,977
lives that had been lost on September 11, but the memorial plaza
would also provide a sanctuary for those who worked in the five
skyscrapers that one day would rise like sentinels around its periph-
ery. Lulled by the soft sound of cascading water, it would provide an

oasis of tranquility and for reflection amidst the downtown hustle and bustle. From the start, I was an enthusiastic supporter of Arad's inspired vision.

But then I watched with dismay as the memorial project began, as *The New York Times* put it with apparent frustration, "spinning out of control." In fact, it was quickly teetering on, the paper feared, "the brink of collapse."

THE BUILDING OF THE memorial was plagued by many problems—and many of them were all too familiar to me.

For starters, there were confrontations with Libeskind, whose master plan had envisioned a subterranean memorial that would not disrupt the flow of pedestrian traffic between the towers. Arad, however, insisted that his two dark pools be placed at street level. They went at it tooth and nail, while the press gleefully reported on the shouting matches between the two strong-willed architects. And having had my own run-ins with Libeskind, and having witnessed what David Childs had gone through while having to work on the final design for Tower One, I could easily imagine the tension.

Arad had disputes, too, with Santiago Calatrava, the superstar architect who designed the wildly expensive transit hub, the Oculus. Calatrava's plan included an extensive series of skylights that would let natural light stream down into the underground plaza of the train station. But since the roof of the Transportation Hub was also the base of the memorial plaza—another example of how the Trade Center site was an intricate puzzle of interconnected parts; a fact that the Port, to my dismay, had never fully appreciated—Arad complained that the skylights would be disruptive, ruining his sober memorial with kitsch. As designed, the skylights were an "incredibly aggressive gesture," the young architect publicly fumed.

Yet to both Arad's and Calatrava's credit, an agreement was eventually reached—but only after more than a year of heated conversations and debate. "I understood that in this particular place, the memorial is more important than the station," said Calatrava. It was a gracious concession from a world-famous architect who time after time had refused to alter his vision for the transit hub. Nevertheless, a young, determined novice architect, after a long struggle, had managed to obtain this victory.

I was impressed by Arad's coup. Good for him, I cheered. Yet I also found myself wondering: If the Port executives had been equally strong-willed, equally determined, would they, too, have been able to rein in Calatrava's grandiosity? Would they have succeeded in saving the taxpayers at least a portion of the Oculus's $2 billion in cost overruns? I could not forget how my team had come up with minimal, nonintrusive ways to alter Calatrava's design that would have resulted in significant savings—and how the Port had simply shrugged off our detailed suggestions. They would rather spend millions than lock horns with the superstar architect.

But with me, they had been combative from the get-go. I often wondered why. Perhaps they had wanted to prove that they were developers, too. That anything I could do, they could do as well or better.

Or maybe it was a cultural difference. They saw me as someone who, as an executive director of the Port put it, was determined "to squeeze the last nickel" out of any deal, while they claimed to represent the public interest. They never appreciated my deep patriotic and civic commitment to rebuilding the site. Sure, I am a businessman. I put up buildings to earn a living. But that was far from the energizing motive in my setting out to bring the World Trade Center complex back to life. I was never able to get them to appreciate that. As for Calatrava, they viewed him as the great artist he is, and that was why, I suspect, they were so easily swayed.

Yet as Arad proceeded, he was abruptly blindsided by security concerns, just as I had been when the New York police objected to both the design and the placement of Tower One. I had had no choice but to scrap the work that had been done and come up with an alternative configuration. Similarly, in 2006, after two years of working on the project, Arad was forced to abandon his plan for belowground memorial walkways because the police feared terrorists might be able to plant bombs surreptitiously in the dark corridors. It would have been an evocative subterranean galley way, the walls of cascading water serving as backdrops for the names of the dead lining the underground passage. But the security assessment was that visitors should remain aboveground.

After all the years he had already spent on the design, this was a powerful setback for Arad. "A tremendous blow," he said. And he wondered, "Do you walk away from a project after a defeat like that? Or do you find a way to come to terms with the new parameters?" In the end, Arad learned, just as I had, that to get something done on the World Trade Center site, you needed to be flexible. You needed to expect the unexpected—and then you had to deal with it when it inevitably occurred.

And not least, Arad had to navigate the roiling emotions of the heartbroken families of the survivors. I, too, had been there. I, too, had to explain time after time to aggrieved people that my intention was not to desecrate the memory of their loved ones, but to create a permanent monument to their sacrifice. In my case, many of the survivors' relatives had vehemently objected to any sort of commercial construction on the site; they preferred that the sixteen acres remain a sort of cemetery in the heart of downtown Manhattan. For Arad, there was a prolonged and contentious debate about the placement of names on the aboveground granite memorial wall. And, same as me, he had to make his case to people suffering through the pain of deep, tormenting emotions. It was an argument that was nearly impossible to win, and the sort of contentious, heartfelt disagreement where, with tempers understandably raw, it was easy to wind up vilified.

Arad had advocated for what he called "meaningful adjacency" on the memorial wall. He wanted the victims who knew one another or had died close to one another to be listed in close proximity. Many of the families, however, argued that this would give some names an unfair prominence. They preferred a more democratic alphabetized listing, or even a random order determined by a sort of lottery.

"For two years nobody talked about anything other than the name arrangement," Arad complained. "There was no fundraising and no progress being made on construction and design."

In fact, by 2006 the cost of the memorial had risen from the original estimate of $350 million to $500 million. And it wasn't just that money wasn't coming in. The entire project had ground to a halt. Just like everything else at the site.

ENTER MIKE BLOOMBERG. WHILE the mayor had no direct power over the Port and for years had to stand pretty much on the sidelines as they mismanaged the rebuilding process, he did have more direct control over the Lower Manhattan Development Corporation supervising the memorial and the 9/11 Museum. In 2006, his frustrations boiled over and he took action.

His first step was to replace the foundation president, who had been in charge of fundraising for the project. At this point, only a little more than $100 million had been collected, a small fraction of what would be needed. The foundation required, he felt, a president with deep pockets as well as ties to the business and philanthropic communities. And he didn't have to look very far to find someone who fit that bill. The mayor personally took charge of the fundraising process and became chairman of the National September 11 Memorial Foundation.

It was an inspired choice. Mike Bloomberg knew seemingly everyone and was very good at calling in favors and deftly twisting arms. When he reached out to me, my partners and I gave $10 million. And the mayor set an impressive example: he personally contributed $15 million (as well as making a $15 million personal loan at only 0.3 percent interest, the lowest rate possible to avoid it being designated a gift. In addition, he donated $100 million or more for the complex's performing arts center). By the time he was done, Mike had raised close to $500 million for the construction of the museum and the memorial.

The mayor and his staff also worked conscientiously to win the support of the surviving families. The city went out of its way to ensure that they had a voice in the process. And, also important, he had successfully negotiated with the Port to make sure that the agency would share some of the contracting costs as well as supervise the memorial's construction.

"The mayor made it real," Chris Ward exalted to the press. "He raised $350 million on his own back, navigated through all the difficulties and the families. He delivered it. Pretty incredible."

I had to agree. It was an impressive performance. And it was a reminder of what could have been done if there had been strong leaders in the offices of the Port or in the governors' mansions, officials who had been determined to get things done, to cut through all the bureaucratic red tape and the politics. It would have made all the difference, saving years of wasted time and billions of dollars.

Nevertheless, the mayor's unwavering commitment to opening the memorial on the tenth anniversary did not come without a cost. Like most of the projects on the site, the memorial was inextricably linked to the adjacent PATH transit hub. The memorial's plaza was part of the Transportation Hub, and the Port was faced with the construction problem of how the plaza would be supported. The hub was years behind schedule and would certainly not be built in time for the tenth anniversary.

The Port's engineering team, now that they were for once being pressed by the mayor, came up with an ingenious solution. They simply reversed the traditional order of construction work. Instead of building the Transportation Hub up from the bottom on Manhattan bedrock, they worked from the top down. That meant that the plaza—and the hub's roof—was built first. The process involved a new design, new engineering blueprints, and a completely new timetable. I have no idea what was ultimately spent, but it must have been expensive as hell. David Samson, the chairman of the agency's board, conceded as much, acknowledging "cost overruns" in the rush to open the memorial on the tenth anniversary. But he refused to provide reporters with the specific dollar amount. But then again, maybe he didn't know. After all, it was not the Port's money. It was just the taxpayers'.

On September 11, 2011, the memorial opened. The overall effect was profound, deeply moving. Mayor Bloomberg, members of the city council and the Port Authority, and many of the 9/11 victims' family members were present. They were joined by President Obama and the First Lady, along with George and Laura Bush.

The visitors passed through metal detectors and walked by security cameras. Then they proceeded through a park (designed by landscape architect Peter Walker) of more than four hundred swamp white oak trees. Finally, they approached the dark pools, the water rushing down into thirty-foot-deep pits before disappearing into

Laura and George Bush, Michelle and Barack Obama dedicate the 9/11 Memorial on its opening on the tenth anniversary of 9/11.

the depths of another, even darker square. It was quite an engineering feat, the veil of water constantly tumbling down the dark-gray granite walls, only to seemingly vanish, and then be recycled. Yet the memorial's most poignant element was the tilted rim of bronze tablets surrounding each pool, which bore the names of the victims. There were nearly three thousand names at waist level, the letters incised into the bronze. Instinctively, people ran their fingers over their loved one's name as if trying to reach out to them one last time. It was heart-wrenching to watch.

One of the visitors, Monica Iken, who had lost her husband, Michael, in the attacks, touchingly told a reporter, "I can go see Michael. He's home."

And although I had never met Michael Iken, I felt in my own way a similar emotion, even a kinship of sorts.

YET WHILE THE MEMORIAL had opened on schedule, the work on the adjacent $1 billion 9/11 Museum dragged on. And once again the reason was not so much money or construction delays, but politics.

Under the original 2006 MDA, the 9/11 Memorial Foundation, which received the land from the Port, was responsible for raising $700 million to build the museum, obtaining the funding from state and federal allocations, as well as private donations. The Port, though, would handle the construction and pay hundreds of millions of dollars in related expenses.

Only in the aftermath of the memorial's opening, work on the museum simply shut down. Fundraising pretty much stopped, and the planned exhibits were left to gather dust in fabrication shops in Buffalo, New York, and Santa Fe, New Mexico.

Governor Andrew Cuomo put the blame on the museum's ballooning costs. The price tag was now estimated to be about $1.3 billion—and, like everything at the site, would be rising, he warned. The Port claimed it had stopped construction because the foundation owed it somewhere close to $300 million. The foundation countered that the agency owed it more than $100 million because it had failed to complete the museum in 2009, as had been originally promised.

I, of course, could understand the foundation's complaint. I, too, had been victimized by the Port's failure to complete construction work as promised. And I, too, had demanded—and succeeded in collecting—hundreds of millions of dollars in damages for these irresponsible delays.

However, the real bone of contention, according to what I was hearing, was not money, but political power. Cuomo wanted more oversight of the museum and foundation. Why? He didn't want Bloomberg, who was chairman of the foundation, to have all the control or receive all the credit. Cuomo wanted the museum to be part of his résumé if he ran for national office.

New Jersey governor Chris Christie had also dug in his heels. He made it clear that he didn't want one additional penny of Port Authority money to make its way into New York. Why? His reasons had little to do with a desire to cut spending. Rather, Christie believed the city had already received plenty of money from the Port. If the agency was going to spend additional sums, he wanted it allocated to New Jersey. Where he would get the credit.

Further exacerbating the raging, self-interested animosities, I had heard from reliable sources, was the governors' belief that they had been slighted by the mayor. Both Christie and Cuomo were fuming that Bloomberg had deliberately, or so they claimed, limited their access at the tenth-anniversary commemoration. When the memorial had opened, they were not placed up front where the television cameras were focused. It was a petty, childish pique. Yet it delayed the opening of the museum and added millions to the final cost. And it was typical of how politics had managed to disrupt the rebuilding process, an insidious fact of life in the world I had been living in ever since I had purchased my leasehold.

Still, this latest political feud, however bewildering, was ultimately merely a backdrop to more consequential events. For in spite
of all the years of squabbles and delays, the towers were finally starting to rise up into the sky. And I couldn't have been happier. My
long, hopeful dream was beginning to come true.

EIGHTEEN

THE SPRING OF 2012 was a heady time. The transit hub had begun to take shape, and cranes were lifting the rafters holding its signature wings into place. More than 4 million people had visited the 9/11 Memorial in the year since its opening. And all you had to do was look up and you could see that the work on the three skyscrapers was proceeding by leaps and bounds. One World Trade Center had climbed to about a thousand feet, on its way to becoming the tallest building in the country. My 4 World Trade Center had risen to about sixty floors, with only another twelve stories to be added. And the podium base of Tower Three housing the PATH terminal power and utilities infrastructure had been built up to the height of an eight-story building. The drawings and models I had first glimpsed years earlier were now at last coming to life before my eyes.

In fact, Tower One and my Tower Four were in a race, competing to see which would be the first to raise the final steel beam to its top. And the Port officials desperately felt they needed to top out first. They hoped a victory would convince skeptics and critics that at last the agency had overcome all its problems.

The Oculus takes shape.

They did indeed need a win. Their ineptitude had finally caught up with them. I had been uncomfortably dealing with their wasteful and capricious ways for years, but now their actions had become embarrassingly public. The previous summer, Governors Cuomo and Christie had initiated an audit of the agency after a big hike in tolls for buses and trains had left commuters fuming. The scathing report issued six months later revealed the costs for constructing the World Trade Center site had ballooned $3.8 billion from the last accounting; the project's price tag was now a bloated $14.8 billion. "Challenged and dysfunctional" was how the audit described the agency. And Governor Christie's reaction was similarly seething. "This is, unfortunately, a testimony to the mismanagement of the Port for years," he railed. "And it's a testimony . . . to their unwillingness to budget effectively."

And so agency officials were looking to change (or at least offer a distraction from) this disparaging narrative. They wanted to give the press another story to report. They hoped to demonstrate that not only had they finally succeeded in getting something built at the site, but they were first to do it.

Me—I had bigger concerns, and bigger challenges, than winning a race. Or a public relations battle. I simply wanted to get my Tower Four completed as soon as possible. Then I could focus on the next one I had to build. And then the one after that. I wanted to get the World Trade Center site up and running.

THERE WAS NO DENYING that Tower One had come a long way from the days of design disputes, construction postponements, and marketing disasters. But it had been a difficult journey. It had taken nine years (at least three more than scheduled) and an incredible expenditure of taxpayer dollars (about $3.9 billion in total, or approximately $1.5 billion more than had originally been budgeted) to get the skyscraper on this path toward completion.

It need not have been this way. In 2006, when I had relinquished control of the building to the agency, I was nevertheless still contractually obligated to construct the tower. In fact, I was getting a fee of $20 million to get it done. The Port, however, didn't want my help. "We'll pay you," the executives told me, "but we're going to do it ourselves."

"What are you talking about?" I challenged. "You're not being realistic. You don't have the wherewithal to get the job done. Your people don't possess either the skills or the competence," I said bluntly. "Let me help you."

The agency officials remained adamant, so I tried a more conciliatory approach.

"At least let my people advise your team," I nearly begged. "We'll make sure you keep on schedule. And when you stay on the timetable, that makes holding to a budget a lot easier, too. You blow your schedule, you blow your budget. And it's in all of our interests," I patiently explained, "to get the site up and running in a timely way."

I was speaking candidly. It wasn't just altruism that was motivating me. For the site to succeed, I had realized from the start, we all had to succeed. All the buildings needed to be up and running. Their delays would affect me, too.

But they didn't want to hear it. They wanted to show that not only could do they do anything that I could, but that they could do it better than someone who had spent a long professional life building and leasing office towers. They didn't need me or my people. They would rather pay me to walk away than get any benefits from the big check I received.

AND HOW DID THE Port's go-it-alone strategy work out?

Well, after four years on their own, they realized they had a problem—and the makings of an even larger one were brewing. It wasn't just that there had been innumerable delays before construction of the 1,776-foot-high tower had begun in earnest, but they were having a hard time finding tenants to occupy the cavernous 3 million square feet of office space. They had only signed up a single private tenant, a Chinese real estate firm that planned to take the sixty-fourth through sixty-ninth floors, about 190,000 square feet. Which left a lot of ghostly empty floors that still needed to be filled.

The chagrined agency realized they were losing control of the project, and for a while they explored selling the building to a private equity firm. The recession, however, had made that sort of deal impossible. So the Port pivoted. They decided instead to find a partner who could help them get the skyscraper completed and who could bring in tenants.

In the summer of 2010, the Port wound up making a deal with the Durst Organization. Durst, a family-run real estate group that owned ten Midtown buildings and whose principals I had known for years, would now supervise the completion of the tower, which was costing an enormously high $1,269 a square foot. And they would also lease and manage the commercial space.

In 2007, the Dursts had contributed to the full-page ads in the New York papers that urged the newly elected Governor Spitzer to put an end to the Freedom Tower. The deal they cut now for their participation was a shrewd one. In return for $100 million, they received a 10 percent equity interest in a tower that had so far cost $3.3 billion. That gave them for starters a potential 200 percent return (at least) on their investment. Yet in addition, they got a $15 million management contract that rewarded the Durst Organization with 75 percent of any monies saved by cutting construction costs up to $12 million, and 50 percent of every dollar saved after that. That is, the Port was now pretty much paying the Dursts millions to do what I had already been paid for—only the agency had told me to get lost.

The Dursts, canny developers, quickly found ways to cut costs. They went to work on the building's large base and found it could be encased with a cheaper version of prismatic glass. And they insisted that the radome—the sculpted shell of fiberglass and steel that would protect the antennae and maintenance platforms on the tower's summit—be abandoned. It was a redesign that saved the

Port about $20 million, while at the same time it earned the Dursts, thanks to the generous deal they had negotiated with the agency, at least half that amount.

The Dursts' most significant initial contribution, though, was to find an anchor tenant for the tower. They convinced Condé Nast—a leading magazine publisher and media company that was occupying space in a Durst-owned building on Forty-Second Street—to move downtown. The Dursts allowed the publisher to cancel their Midtown lease, and in return Condé Nast agreed to a twenty-five-year lease on more than 1 million square feet—floors twenty through forty-four—in Tower One. This was a tremendous boost not just for the prospects of the building, but for the entire site. It demonstrated to the world that the World Trade Center complex was going to be a viable development, that employers would want to move their businesses to Lower Manhattan. It would help attract other tenants, and not just to Tower One, but also to my buildings. Success, as I have always said, breeds success. It was a win-win for both the Port and me.

But Tower One still had to be completed. My big fear was that if the Port didn't finish the infrastructure needed to get the site up and running, then even the most game-changing lease agreements would become irrelevant. Condé Nast needed a firm date when they could move in. They wouldn't tolerate the Port's mercurial timetables. They needed to make plans for a move on a specific date. I grew anxious that Condé Nast would become infuriated by the agency's failure to deliver the floors as promised and then, with good reason, rip up their letter of intent.

This concern was further fueled after I talked to the steel contractor who was erecting my Tower Four. He was also doing the same job at Tower One, but, he complained, the Port was tens of millions of dollars behind on payments that were due. He said the agency refused to pay because they were unhappy with the work.

This left me perplexed. Silverstein Properties never had any problems with his work. And we were always right on time with our payments. But, I was learning, we did business differently than the Port.

And I worried that their contentious posture would force the steel contractor to shut down until they paid up. If that happened, then Tower One would be further delayed as charges and counter-charges would inevitably be hurled. It would likely end in litigation, the latest in a long string of lawsuits that could only sour potential tenants' assessment of the entire site.

My thoughts only grew more anxious after I talked with Doug Durst, the president of the company that was now running things at Tower One. He told me that his decision to work with the Port had been a tragic mistake. It didn't matter how much money he stood to make, he said. "It's just an impossible group of people. There's a complete lack of competence. How did you manage to work with them for so long?" he asked.

In fact, he told me, he had wanted to break his deal. It hadn't taken him long to realize he needed to cut ties, and so he had gone to Chris Ward, the agency director with whom he had negotiated his partnership arrangement. "I got to get out," he told Ward. Ward, he said, had pleaded with him to stay. "It'll get better, I'll help you," Ward had promised.

But Ward never got the chance to make good on his vow. Cuomo decided that Ward should be the fall guy for the Port's mishandling of the Trade Center. Ward was replaced by Pat Foye, and Durst, worn down, resigned himself to staying on.

Which was fine with me. I was counting on Doug Durst to get the tower finished on a schedule that would make sure Condé Nast didn't have any second thoughts about moving in. The presence of a major media company in the complex would make it easier for me to attract tenants at Tower Four. And if the Port wanted to make this into a race to see who would top out first, that was fine with me, too. I didn't care too much about who won. I wanted the complex moving forward. The topping-out of Tower One would be a victory for me, too.

BUT MY TOWER FOUR was indeed rising. And I already had a good head start in getting tenants. Under the terms of the 2010 financial backstopping deal I had made, the Port had agreed to lease 600,000 square feet, and the city had committed to an equal amount. That meant that out of 2.3 million square feet of office space in Tower Four, I started out with guarantees for a bit more than half. I was sure I would be able to lease out the rest of the building.

So we kept building, adding about a floor a week as we headed up to the summit, seventy-two floors high. We were right on schedule and right on budget. Things were going swimmingly. But then in February 2012, a crane was lifting almost twenty tons of steel to the forty-first floor when the cable snapped.

The entire load of steel crashed down forty feet, demolishing a flatbed truck. Miracle of miracles, no one was hurt. Fortunately, the driver of the truck had followed the standard safety protocols and had left the cab while the steel was being raised from the truck bed. If he had not, he never would have survived. Building a skyscraper was an enterprise that was filled with dangers, and part of a developer's responsibility was to make sure all the safety precautions were obeyed. Thank god they were on Tower Four.

Still, the crane was no longer functional. It had to be deconstructed, taken down from the top of the building. That meant we now had only a single operating crane. With two cranes, we could do a floor a week, sometimes even a floor and a half. A single crane cut down progress considerably.

For a while, though, we tried to make do with just the one. The steel contractor, who was really the person most responsible for the pace and schedule of construction, had his crews working overtime to try to approximate the job that two cranes could do. The men were doing double, even triple shifts to get the steel raised. But it still wasn't fast enough. The contractor, at my urging, finally decided to get permission from the Port and the city to erect a second crane on the roof.

Only once that was done, the rain started. There were buckets of

An ironworker leans out from the top of One World Trade Center in 2012.

rain and terrible, howling winds for days. In those kind of conditions, you just cannot raise steel. We lost maybe another five days because of the weather.

But as spring arrived, we were back in business. Plenty of sunshine and two cranes operating. It looked like we would top out in June, ahead of Tower One. The Port had started construction about two years before us, but we would be first to raise the final steel beam to the roof. We would win the race. But, of course, who was keeping score?

I soon found out.

"HOW WOULD YOU FEEL about a joint topping-off ceremony?" Pat Foye, the newly appointed Port Authority chairman, asked in a call one morning in April. "Would you want to talk about it?"

I was puzzled. It was clear that my Tower Four would be completed before Tower One, so at first I didn't understand what he was suggesting. But then he explained.

"The thing is," he went on sheepishly, "it'd be embarrassing to the Port if your building topped out first."

I could see his point. We had started construction two years after the agency. Now we would be raising the final steel beam to the roof of Tower Four before the steel would be put into place on the top of their tower. Plus we had brought our building in on budget, $1.67 billion as promised. It was a seventy-two-story environmentally sophisticated LEED-Gold-rated building designed by Pritzker Prize–winning architect Fumihiko Maki. A refined, carefully detailed skyscraper that rose with a somber, dignified grace above the two dark pools of the 9/11 Memorial. And now we would be beating them to the headlines. It would be the first building to top out on the new World Trade Center site. This wouldn't help the agency's already dubious reputation, recently further tarnished by the searing audit report, one bit.

One part of me wanted to say, "Serves you right. After all the pain the Port has caused me, after all the millions I had to spend on litigation just to get a fair shake from the agency, it would be good to have a long last laugh."

But I also knew this was business and I needed to push emotions aside and do whatever was in the best interests of achieving my goal. And my goal, same as it had been for the past ten and a half years, was to bring the World Trade Center complex back to life. I couldn't accomplish that without the Port. Our futures at this site were intertwined; any success they had would translate into good news for Silverstein Properties and our towers, too.

"My god," I told Foye, "if you want to do a joint topping-out, let's do a joint."

When I got off the phone, I sent word to my team supervising the construction at Tower Four. "We need to delay putting down the final rooftop beam," I ordered. "Everything else—full speed ahead."

TWO MONTHS LATER, FOYE called again. "There's been a change of plans," he began. But he hesitated before continuing. Clearly, whatever he was about to say left him uncomfortable.

I just waited, not knowing what to expect. But whatever it was, I suspected I wouldn't like it.

Finally, he spoke. "There's no longer going to be a joint topping-out. We're going to go first."

I was blindsided, and stunned. "Pat, we had a deal."

"Yes," he agreed. "But a new decision has been made upstairs. By people with a higher pay grade than mine."

"I had assumed," I said pointedly, "when you proposed the joint ceremony, you had full authorization."

"Things always change," he countered with what struck me as thin conviction.

"No, they don't," I shot back with force. "You make a deal, you stay with the deal. And we had a deal."

"I can't do it anymore." His tone was flat and resigned.

I almost felt sorry for him; he had been dispatched on a very embarrassing mission. "What are you proposing instead?" I asked.

He explained that the Port would be ready to top out Tower One in about two weeks. After they got that done, then I could have my ceremony for Tower Four a few weeks later.

"Pat," I began evenly, determined not to lose my temper. "We're not going to do that."

I paused. I wanted to make sure he understood what I had said. And that he grasped my resolve.

Then I proceeded with what I hoped he would see as a conciliatory proposal. "As far as I'm concerned, you want to go together, we'll go together. But if not," I continued, "we're not going to wait. We're ready to go. We can go tomorrow, or, worst case, two days from now. But if we're not doing this together, there's no need for us to wait weeks. We'll go soon."

"You can't."

"*What do you mean, I can't?*" So much for my decision to keep calm.

His response was both cryptic and ominous at the same time. "There are people who don't want you to go ahead of us."

By now I had heard enough. "Pat, that's not going to happen," I said with a bristling resolve. "Look," I went on, "there's nothing further to discuss. We're prepared to honor the deal we'd made with you. If you don't want to go ahead with that arrangement, then we'll proceed on our own. And you should know—we can have the topping-out tomorrow."

"You can't," he repeated once more, and this time with even more desperation. "The Port will stop you. They'll find a reason to stop you."

I felt like I was in a scene in a mob movie. I was being presented with an offer I couldn't refuse. Would I find a severed horse's head in my bed one morning?

"Pat," I tried, "this makes no sense at all. The Port can't stop me. It's my building."

"Larry," he warned once more, "they will stop you. The Port will find a reason to stop you."

"Look, Pat, we're going to have our topping-out. You deliver that answer to whomever you have to. And let them know that the Port's invited to the ceremony. You come, and everyone else can come, too. Let's all of us participate as if everything is fine. As if there's no rift of any kind."

"Larry, this is a disaster," he said plaintively.

"No, it's not," I responded in an attempt at reason. "There's nobody foolish enough to let this end in disaster. Not at this stage. Not when we're so close after all these years to being able to fill our two buildings with tenants."

But Foye's only response was to repeat that Tower Four could not top out before Tower One. The Port would look foolish.

By this point, I had come to think the entire conversation was foolish. I had had enough. "You do what you want, Pat," I said. "But you're on notice. We're going to top out very soon."

And with that promise, I hung up.

Our topping-out was set for June 25, and all the invitations had been sent. We had invited the mayor and other politicians as well as many of the construction union officials. The newspaper press and the TV crews had indicated they would be coming, too. And just a few days before the ceremony, I got a call from Foye.

"Okay," he began, "we're not going to stand in your way."

Stand in my way? I thought. *What could they do? Dispatch a bunch of thugs to block the entrance to my building?*

However, I kept my snide thoughts to myself and, instead, tried to be gracious. "Pat," I said, "you and your colleagues are invited. In fact, we'd like you to speak. You'd be an honored guest. The Port is our partner."

I continued on, eager to persuade him that it was in all our inter-

The topping-out ceremony for 4 World Trade Center on June 25, 2012

ests to present a united front. "And I want to guarantee you that we won't mention we are first. We won't say anything negative about the Port. We want people to see that the Port and Silverstein Properties are working together to get the Trade Center up and running."

"Okay, we'll come," he finally said.

But he never did. Nor did anyone from the Port.

Nevertheless, the ceremony on June 25, 2012, was a big success. The final steel beam was signed by more than one hundred construction workers before it was placed, along with an American flag, on top of the 977-foot tower. John Rzeznik, one of the project managers at the site, spoke. "Everybody's put their blood, sweat, and tears into this," he said with emotion.

I knew precisely what he meant. I had lived through a long, demanding eleven years to get to this point. But I also knew my troubles had been minimal when compared to the price that all too many construction workers had to pay. There had been, it was recorded, thirty-four significant accidents over the years as the buildings rose up. And while this was not an alarming statistic for a project of this scope, the specifics were harrowing. A few of the incidents, in fact, had been permanently disabling: a worker fell twenty feet when a scaffold collapsed; another was struck in the head by a sixty-pound bundle of rebar.

And tragically there were two deaths. Hugo Martinez, thirty-six, an employee of L&L Painting of Long Island, had been crushed by an aerial lift in June 2003. Then, in 2004, a twenty-eight-year-old apprentice carpenter fell through an unsecured floor hole cover (the reason for this oversight was never resolved) in my Tower Seven to his death.

Working on the exposed beams and the skeleton floors of a tower rising high into the sky was always a dangerous occupation, but in October 2012, safety concerns required additional prudence. Super-

Superstorm Sandy floods part of the World Trade Center construction site on October 29, 2012.

storm Sandy with torrents of rain and gusts of eighty-mile-per-hour winds pounded downtown Manhattan, and all work prudently came to a halt. It was nearly a week before it would resume. Huge pumps had to be brought in to drain the millions of gallons of floodwater— ten feet deep in some places—from the construction sites. And in their deep basement depths, fish that had been lifted by the storm from the nearby Hudson were found swimming.

YET EVEN IN THE aftermath of the ceremony at Tower Four, I continued to be perplexed by what had occurred with the Port. Who specifically had wanted the joint topping-out? And on whose authority had that plan been jettisoned, only to be replaced by a new edict insisting that their building needed to go first, on its own? Even more mystifying, who had authorized the tough-guy threats, trying to tell me what I could or couldn't do at my own building? What were they trying to accomplish? And why, in fact, had they even raised the possibility of a joint ceremony when their tower was not even close to being completed?

As it happened, a couple of weeks after the ceremony I had a chance to talk to David Samson, who had been appointed chairman of the Port's board by New Jersey's Governor Christie. I had known him professionally for years; he had been a distinguished partner in a prestigious New Jersey law firm (although years later he would be forced to leave the Port in scandal, and he would spend a year confined to house arrest for trying to use his influence to restore direct flights from Newark Airport to his South Carolina vacation home). In fact, he had previously represented us on matters relevant to the World Trade Center. And so I felt I could be direct with him.

"Listen," I said, "can you tell me what all that was about? The Port and my topping-out ceremony?"

"What are you referring to?" he asked.

It was clear he had no idea, and so I went through everything.

The broken agreements. The bizarre threats. The refusal to come to the event.

"Larry," he said, "I never heard any of this before. I never had a clue. And I can tell you, it didn't come from anyone in New Jersey. If it came from the governor or any of his appointees, I would've heard about it."

"But why the hell would it have come from anyone in New York?" I wondered. "The mayor, the governor—they would've reached out to me directly if they had a problem."

"Larry," he continued, "you have no idea how strange things are at the Port. I've been there for about a year now and let me tell you, it's much worse than anything you'd ever described to me. The level of dysfunction. The arrogance. They live in their own world. What I've seen, it's an eye-opener."

As for who had issued the order that Tower One needed to top out first on its own, and as for who had thought it would be a good idea to try to intimidate me, well, that still remains a mystery. I never found someone who could give me a straight answer. It continued to be a totally bewildering experience. Yet only one of many in my years of working—or, more precisely, trying to work—with the Port.

What was a certainty, however, was that Tower Four opened on November 13, 2013. And that Tower One opened nearly a year later, on November 3, 2014. And with those two events, at last the Trade Center site, thirteen long years after it had been destroyed by terrorists, was up and running.

One World Trade Center and 7 World Trade Center, with the 9/11 Memorial
Museum in the foreground

NINETEEN

I ALREADY HAD A HEAD start on Tower Three. The Port, once again acquiescing to the costly demands of Calatrava's unyielding design, had agreed to house the lion's share of the transit hub's infrastructure not in the Oculus itself, but off-site in the base of my Tower Three. This quixotic decision required that I build the tower up to the eighth floor to fully accommodate these mechanical workings, and, even better, I was able to do it with some significant financial support from the agency. In fact, we came out way ahead on the deal. The Port, once again cavalierly spending commuter revenues, wound up paying for things, including much of the building's lower-floor retail store space; in a normal project, this would never have been covered by a second party. However, constructing the commercial office tower that would rise up above this podium base, the Richard Rogers–designed, eighty-story, $2.75 billion skyscraper—the fifth tallest building in the city—turned out to be much more problematic.

At first, though, I had thought with my usual optimism that

things would be well on their way after the hard-fought deal I had negotiated with the Port back in 2010. That arrangement provided up to $600 million in backstopped financing ($390 million from the agency, the city, and New York State, as well as another $210 million from the city and state). But stern benefactors, they made sure the gifts came with long strings attached: the backstop financing would only be provided if I put $300 million of cash into the project, secured leases for about 400,000 square feet (that is, for a little less than a sixth of the tower's 2.5 million square feet of commercial space), and raised the remainder of the financing I would need to get the multibillion-dollar skyscraper completed on my own.

Still, I thought, these were goals we would be able to fulfill. We would swiftly find the anchor tenant we needed and that, in turn, would pave the way for our raising the financing. At least that was my wishful plan.

My confidence was grounded in what we had to offer to prospective occupants. It was a very appealing package. For starters, Rogers's design was not just visually stunning with its distinctive K-braces, but it also allowed for wonderful office spaces. The big floors were virtually column-free, and that would attract the financial community, which coveted large open spaces for their trading floors. And with thirteen-and-a-half-foot ceilings and walls of floor-to-ceiling glass, you could look up from your desk and stare out at some of the greatest views on the planet. Plus, the building would have all the latest high-tech and environmentally sustainable bells and whistles, and there would even be a 5,000-square-foot terrace on the seventeenth floor where the office workers could unwind, with a panorama of the Hudson River and the spires of downtown New York as the backdrop. And the rents, as I have mentioned before when pointing out some of the economic reasons I had been so eager to build on the World Trade Center site, came with several built-in tax and energy incentives, since the Port, a quasi-governmental agency, owned the land. These perks ensured that renters would be getting not just a good deal but a uniquely advantageous one; that is, they would get a lot of bang for not too many bucks, especially when compared to Midtown office prices.

And pretty quickly it looked like I had been right.

IT WAS NOT LONG after we had started constructing the tower's podium in 2010 (heading toward a 2013 completion date for this base) when it seemed we had landed our anchor tenant.

UBS, a multinational investment bank and financial services firm, started talking with us in earnest. It was just the sort of space they needed; the chance to have massive, open trading floors had, as I had predicted, sold them. After about a year and a half of very detailed negotiations, they were ready to sign a lease for 1.15 million square feet. Which was just terrific. It would not only fulfill one of the requirements the Port had demanded for the $600 million in backstopped financing to kick in, but it would make it easier for me to go off into the financial marketplace and raise the additional $1.5 billion I would need to erect the seventy-three stories that would rise up from the base.

Only just as the lease was being finalized by the lawyers, in early September 2011, the bank issued an absolutely terrible earnings report; profits had nosedived 70 percent over the past year. And if that wasn't bad enough, about two weeks later they reported a $2.4 billion trading loss. When my assistant told me the bank's CEO was on the line, I didn't want to pick up the phone. I could well imagine why he had called.

"Larry," I heard him say, "we'd love to sign this lease. We need the space. But I can't sign. Not with what we've gone through lately." He assured me, though, that if the bank's prospects improved, he would call me back.

I knew better than to wait by the phone. We had been having conversations, all a lot more preliminary than we had had with UBS, with other potential renters from the financial community. They, too, had been initially attracted by the prospect of consolidating their trading departments in the tower's large, column-free floors. Only their interest had also recently fallen by the wayside. In these uncertain economic times, financial concerns weren't confident about making plans that would jump them forward into the future. Rather, they were prudently sitting back. They wanted to wait, not take on any new obligations, and see if the economy improved. Each and every one of the firms to whom we had been talking decided to stay where they were; they wanted to see where things stood in about five years. Which meant that, in the best of all economic futures, we would not be able to conclude a lease with any major player in the financial services industry for probably another seven years. And that meant Tower Three would not rise up higher than its base for quite a while.

UNLESS I COULD FIND another sort of tenant. And that, I told myself, was the good thing about New York. Sure, financial concerns had dominated Lower Manhattan for decades (and decades), but this was a new era. There were a host of fresh, affluent companies, the latest tech, communications, and media concerns, thriving in New York. They were the sort of rock-solid companies that could lease an entire $2 billion, 2-million-square-foot building just for themselves. After all, Google had done precisely that.

And just as I was having these hopeful thoughts, along came GroupM. Actually, they didn't just come along. They were brought to us by Mary Ann Tighe, the commercial real estate broker who had also been instrumental in convincing Condé Nast to move to One World Trade Center. Mary Ann, who was the chief executive officer of the New York Tri-State Region of CBRE, the world's largest commercial services firm, was a powerhouse, an art historian by training who had become an innovator in what had typically been a male-dominated industry. She became an invaluable ally as the Trade Center complex continued to take shape.

And she had made a brilliant match when she introduced us to GroupM. They were a world-leading media investment and advertising concern with several Midtown locations; all told, they employed about 3,500 workers throughout Manhattan. It would make better sense, they decided, to consolidate 2,400 of those employees into one building; staff interaction could, they felt, enhance creativity and productivity. The company executives took a good look at the massive open-space trading floors that had been designed for Tower Three and decided this was precisely the sort of layout they required. It would allow them to have a synergistic work area, the employees interacting and feeding off one another's ideas. Originally, they came to us wanting about 715,000 square feet in the building, but they realized that the open, column-free expanse had been designed so efficiently that there was no need for so much space. All their employees could be comfortably located on nine floors of Tower Three, about 516,000 square feet.

That was less space than they had originally wanted, but I was not complaining. In fact, I was thinking that even if this wasn't a home run, it was a solid triple. I would have the anchor tenant that would allow me to go out into the marketplace and get the financing I needed to build. Only I had to start construction quickly. GroupM's uptown leases would be expiring over the next three years, and if Tower Three wasn't ready for occupancy, then they would nullify their agreement. And who could blame them? If I couldn't deliver the building, they had every right to walk.

So I got to work. I conferred with the bankers at Goldman Sachs and JPMorgan Chase I had brought on board to obtain the construction financing we needed, and they were enthusiastic about the prospects. Mortgage rates had gotten so cheap that they were confident that even with a single anchor tenant taking just a half-million or so square feet, they could get the money I needed to build up from the podium. And, the icing on the cake, I could get the approximately $1.4 billion in financing at a very attractive rate.

So, I went full speed ahead. I shook hands with GroupM. We got a deal, I told them. And it didn't take us long after that to sign a memorandum of understanding. Now I just needed the bankers to do their part. I needed them to deliver the financing they had so confidently assured me they would be able to raise.

However, the process dragged on and on. But I was not anxious. At least not yet. I figured that raising almost a billion and a half dollars had to take some time.

But then I started tracking a series of ominous events. First I read in the papers that the Fed was thinking about cutting back on its purchases of Treasury bills. Next, I saw that T-bill rates jumped from 1.5 percent to 3 percent in four days. They had doubled! And just like that, the spreads the banks were charging above the T-bill rate went up about 100 basis points, too.

So much for my having to pay only a meager 1.5 percent interest on the billion dollars or so I wanted to borrow. That world no longer existed. In a flash, the mortgage market had completely changed— and not for the better for a borrower like me. The cost of the money I had hoped to borrow had more than doubled; the rate for loans was now 3.5 percent. I could no longer afford to finance a $1.4 billion loan. At this new rate, and with only a single tenant paying rent, I wouldn't be able to service my debt.

And if I couldn't obtain the construction financing I needed, I would be unable to build. I worried about how long GroupM would wait before finding another location. The company had leases expiring. It needed offices for their workers. If I couldn't provide the space in Tower Three, I had no doubt they would find another building that could.

There was a simple solution to my dilemma: get the financing I needed to build. But how? Then I had an idea. Actually, it was an old idea. It had worked before, and now I wanted to believe it could work again.

I WENT TO THE Port and asked them to backstop the financing with the agency's credit. Their guarantee, along with my having GroupM as an anchor tenant, would allow me to raise the $1.4 billion I needed to begin construction. And I carefully pointed out how their working with me would benefit the Port.

For one thing, it would get another tower built on the complex. That would add to the momentum, the sense of ongoing activity at the entire site. It would be good for Tower One, and for all of Lower Manhattan, to have the sixteen acres hustling and bustling with office workers. It would show that the complex was up and running, a viable location for businesses. And if for some reason I failed to service my debt and wound up defaulting, well, then the Port would get title to an eighty-story, 2.8-million-square-foot office tower at no substantial cost.

I also pointed out that once Tower Three was up and running, it would mean that the retail stores in the building would be open, too. This would bring in, thanks to the Port's deal with Westfield, which had taken over operation of the entire retail component of the Trade Center, about $100 million in annual fees to the agency. In addition, the Port would receive another $100 million or so of ancillary fees that would be due upon completion of the skyscraper.

Consider the alternative, I argued. If Tower Three stood uncompleted on Greenwich Street, a podium without its tower, this would cast a pall over the entire site. It would help perpetuate the public's perception that the complex would never be finished. And that would be a disaster for both the agency and Silverstein Properties. Would the Port, I pointedly challenged, be able to find additional tenants for its Tower One in such a climate?

The Port executives seemed interested. But too much was at stake; I couldn't just sit back and hope the board would vote to approve a backstopping deal for Tower Three at its next meeting. I needed to make sure they wouldn't change their minds. In business, it was always a mistake to leave things to chance. And the Port executives, experience had taught me, were a very mercurial group.

I went to Governor Cuomo to ask for his support. How do you get people on your side? You need to show them how they will profit from what you are suggesting. They need to understand what is in it for them.

So I began connecting the dots for the governor. Building Tower Three, I pointed out, would provide work for seven thousand or perhaps even eight thousand construction workers. That meant a lot of families would be getting paychecks. And construction workers famously spent their paychecks, I went on, where they lived and worked. That would bring in hundreds of millions in sales as well as income tax revenues to the state and the city. It would be a win for the workers and a win for local government. I just need your help in convincing the Port to backstop the financing, I concluded.

"I get it," the governor said. "I'd like to see this done." And the governor's churning mind saw an advantage I had not.

What particularly excited Cuomo was the possibility that a deal could be fortuitously timed. He wanted to announce the resumption of construction on Tower Three just as he would start his re-election campaign. He would be able to promise voters thousands of new jobs now, as well as the prospect of a new building at the Trade Center being completed during his second term. He hoped to look like the can-do dealmaker who had rushed into the fray and resurrected a stalled project. He wanted the credit for getting Tower Three built.

I had no problem with that—as long as he was able to use his political clout to get the Port to approve the $1.4 billion backstop financing I needed.

BEST OF ALL, GOVERNOR Cuomo had a plan. Or, more precisely, a plan to get us a plan.

"I've got just the guy who can get this done for you," Governor

Cuomo told me. "He'll be able to deliver the Port," the governor assured me. "You'll get the financing you need."

And you will get the re-election push you are hoping for, I thought to myself. But, of course, I knew better than to say that. Nor did I mind his winning any votes from his support for my eighty-story skyscraper. If Cuomo could help me get Tower Three built, I would be all in favor of his serving another term as governor. And, for that matter, another after that.

It wasn't long after this promising conversation that I was put in touch with Scott Rechler. Rechler was vice chairman of the Port's Board of Commissioners, but the impressive title was a bit of a façade. He had been appointed to the authority's executive committee by Cuomo, and his job boiled down to a single duty—to do the governor's bidding.

I hit it off with Rechler right away. He was a businessman, and in my book that was always a compliment; it meant he had a practical mind and a can-do attitude. It didn't take him long to grasp the logic behind what I wanted from the Port, or to understand why it would also make good economic sense for the agency. And he made the Port's potential deal even better. He got me to agree to pay hundreds of millions of dollars of fees that, without the backstop financing, would have been the Port's responsibility. He was an astute negotiator, and he delivered a real windfall for the agency—if he could get the deal finalized. It would cost me, but I had to admire his advocacy for the Port.

Starting in August 2013, Rechler threw himself into the task of getting all the commissioners to support the financing plan. His first step was to persuade the agency's group of New York directors, and he quickly made good progress. This was, of course, not very difficult; these board members were appointed by Governor Cuomo and were by instinct (as well as self-interest) sympathetic to his wishes. As for the directors appointed by New Jersey's Governor Christie, persuading them would be more challenging; Rechler clearly lacked the leverage he had with the New York contingent. But that didn't stop him from trying. It was an impressive, hard-driving campaign.

And all the while Rechler was pushing and cajoling, I was moving forward with GroupM. We signed a provisional twenty-year lease with the company in November 2013. They would occupy 516,000 square feet on nine floors. Here was the anchor tenant I had needed.

At the time, confident that Rechler would get the job done, I told the GroupM executives that the Port should approve the backstop plan for our construction financing by the end of the year. Worst case, I added candidly, the issue would not be resolved till the first quarter of 2014. But certainly no later than that. By early next year, I pledged, we would be well on the way to securing the billion and a half dollars we needed to start building the tower that would be their new corporate home.

Only no sooner had I offered up this timetable than I began to fear that I had been overly optimistic. *Had I promised too much?* I suddenly worried. Because to my chagrin, a sheepish Rechler now informed me that he was having more trouble than he had anticipated in getting the New Jersey commissioners to support the deal. The chief antagonist, he said, was Pat Foye, the executive director of the Port. Foye wanted Tower One to be open and fully leased before any new buildings opened in the complex. He didn't want the Port's skyscraper to be competing with another tower on the site for tenants.

Scott deftly did his best to deflate this concern. "It will be three years before Tower Three will be finished," he pointed out to Foye. "By that time, Tower One will be open and leased. Besides," he added, "it will be a huge advantage in attracting tenants for Tower One if they know GroupM is also coming to the site. It'll make the entire development seem more vibrant. A place to be." He also reiterated the points I had previously made about the large pots of money— $100 million for starters—the Port would receive once the retail stores in all three towers were fully operational, as well as the hundreds of millions in fees Silverstein Properties would now have to pay to the agency under the new deal.

But there was another bubbling background concern: Foye had a coterie of supporters among the New Jersey commissioners. Their arguments were, well, more personal. It seemed they weren't so much objecting to the Port's guaranteeing the loan or to the presence of another tower on the site competing for tenants. It was simply that they didn't like me.

What had I done? I wondered. *And how could I make amends?* I had my friend New Jersey senator Frank Lautenberg reach out to one of the hostile commissioners to see what was at the root of the man's animosity. Had I inadvertently offended this guy? I hoped Lautenberg, who certainly was a well-liked and respected figure in New Jersey, could get to the bottom of this mystery. I wasn't trying to win a popularity contest; I just wanted to get my tower built.

"Silverstein, I won't talk to him," the angry commissioner told the senator.

"Why?"

"I don't know. I just won't talk to him," he insisted adamantly.

But Senator Lautenberg kept pushing. "You ever talk to him?" he asked.

"No," the man conceded.

"So why all this anger?" the senator tried.

"I don't want to discuss it," growled the commissioner.

But in spite of Foye's objections and the bewildering personal antagonism toward me, Rechler in time succeeded in lining up the votes he needed. The commissioners ultimately were persuaded that building Tower Three would be a boon for the entire World Trade Center complex and, therefore, for the Port. Sure, things had not gone as smoothly or as quickly as I had originally hoped, but I now felt sanguine that the board would at last vote to approve the financing deal at its April 2014 meeting.

But, as was so often the case when dealing with politicians, some-

thing totally unexpected took control of the situation. It didn't matter that the event had really nothing at all to do with the World Trade
Center. It still knocked all my plans for a loop.

ON SEPTEMBER 9, 2013, and for several days thereafter, two of the
three lanes at the main toll plaza for the upper level of the George
Washington Bridge were shut down during the morning rush hour.
Cars were backed up for miles. The gridlock on local streets brought
all traffic to a standstill. Ambulances couldn't respond to 911 calls.
Things were so bad that the adjacent town of Fort Lee declared "a
threat to public safety."

What had happened? Why was rush hour traffic shut down without warning? What was the emergency that had created such chaos?

The answer, according to press reports, was politics. Or, more
precisely, grudge politics. The lane closures had been ordered, these
articles suggested, by members of Governor Christie's staff to punish
a local Democratic mayor who had not supported the Republican
governor in the 2013 election. As a result, the U.S. attorney for New
Jersey launched a federal investigation that would ultimately indict
three Christie staff members as well as, on a charge unrelated to the
closures, David Samson, the chairman of the Port Authority board.
(Samson would resign his chairmanship; one staff member would
plead guilty; and the other two were convicted in jury trials, but the
convictions were ultimately dismissed by the U.S. Supreme Court.)
And the front-page controversy would also put a damper on Christie's presidential ambitions. The press, inevitably, called the scandal
Bridgegate.

The George Washington Bridge traffic jam brought things to a
halt at the Port, too. It seemed as if every commissioner, every executive was warily looking over his shoulder to see if he was in the U.S.
attorney's crosshairs. The Port, a multibillion-dollar agency that ran
airports, bridges, and tunnels, had, other than for its necessary day-
to-day operations, simply shut down.

And the vote on backstopping my financing? It was postponed, and then postponed again. There wasn't even anyone Rechler could approach. It seemed as if everyone was too busy wondering who would lose their job. Or, even more unnerving, who would go to jail.

And as for Governor Christie, while I originally felt I had his support for a bond issue guaranteed by the Port's finances, he now panicked. He was no longer willing to take a public stand. He couldn't risk making waves. It seemed to me he was fearful that he would be dragged into the Bridgegate federal investigation as a target, perhaps even named as the co-conspirator who had given the order for the vindictive shutdown. He decided that at this politically precarious time it would be more prudent to do nothing rather than make a decision involving a billion-dollar bond issue. He didn't want his name in another news story that would give his critics a fresh chance to go after him.

Christie's cautious change of heart, I was told, had the effect of undermining Cuomo's commitment, too. He also abruptly decided it would be more judicious to keep his head down. A political storm was raging across the Hudson in New Jersey, and Cuomo worried—or at least this was the message delivered to me by someone close to the governor—that if he made a rash move, it could very easily spread to New York. His previous enthusiasm for guaranteeing a bond issue to build an office tower had dissipated. He now feared it might cost him votes.

With that, all the support Rechler and I had carefully assembled for the backstopping had become fragile, if not entirely unreliable. I was starting to panic. Then things got worse.

A RENEGADE NEW YORK Port commissioner had begun speaking to the press. In the entire history of the Port, that had rarely happened. The rules of behavior for board members, while never written down on paper, were sacrosanct: You were appointed by the governor, and therefore you did what the governor wanted. Your job was, largely, to make sure the governor's agenda was accomplished. And the one thing you would never ever do was go to the press. You did not hunt down reporters and tell them that the governor was wrong.

But this was precisely what Ken Lipper did. Lipper, after a career in the financial services industry that had its heady ups and, more recently, some very embarrassing downs, as well as a stint as the deputy mayor for economic development, was a recent Cuomo appointment to the Port board. And now Lipper, although new to the job, decided to make a series of bombastic comments to reporters criticizing the backstopping plan.

He had all sorts of reasons that didn't make much economic sense. He railed about exposing the Port's balance sheet, about whether there already was too much office space in Lower Manhattan. But he never focused on the big picture—turning the World Trade Center site into a viable commercial complex, and the financial return that would bring to the region. Or the hundreds of millions in revenue the Port would receive once Tower Three was opened. He was willing to let the building sit there as it was, a seven-floor-high podium. He didn't care if the World Trade Center was ever up and running. He didn't seem concerned about bringing downtown New York back to life in the aftermath of the terrorist attacks.

What did he care about? You ask me, I think it was more about getting his name in the papers than any economic judgments. My suspicion was that he was simply trying to polish his recently tarnished professional reputation. But maybe I was wrong. Maybe I was being too hard on Lipper. What I do know for sure, though, was that his public criticism, coupled with the political fallout from the Bridgegate fiasco, upended my desperation strategy for getting Tower Three built. The Port's board rejected my request to guarantee my construction financing.

I now feared that Tower Three would never get finished.

.   .   .

BUT I DIDN'T GIVE up. And, as fate would have it, in the end I didn't need the Port's backstop. The bond market changed completely, and this time to my advantage.

In late October 2014, after having waited on the sidelines for a year as I had struggled to get the Port's help, I went into the bond market. And now the rates were working in my favor. Silverstein Properties sold $1.6 billion in tax-exempt Liberty Bonds (these investment vehicles were part of a federal economic package to aid New York's recovery from the September 11 attacks). The bonds were backed by rents and secured by a mortgage on the property; the Port didn't have to guarantee a single dollar. It was the largest-ever unrated bond deal in the municipal market. And this deal had been preceded by a $340 million round of financing through federal Recovery Zone Bonds that New York senator Chuck Schumer, who had been zealously committed from the start to the rebuilding, had worked ceaselessly to arrange.

I now, at last, had my financing. I could begin to build Tower Three up from its podium base.

THERE WERE TWO TOPPING-OUT ceremonies. This was because Tower Three was constructed using a technique that, because of the expense, had rarely been employed in New York: it was climbing up into the sky from the inside out.

Specifically, the skyscraper's concrete core rose up ahead of the perimeter steel columns and beams. And these concrete walls were three feet thick and embedded with steel rebar. They would be a stabilizing force, a spine that would keep the tower steady on a windy day. Or make the building, for all practical purposes, impervious to a terrorist attack, and therefore it was money I felt had been well spent.

On June 23, 2016, a bucket decorated with the signatures of the construction workers was hoisted to the top of this thick concrete core. Four months later a more traditional topping-out took place as the last steel beam was raised to the eightieth floor.

Topping-out ceremony for 3 World Trade Center on October 6, 2016

Marty Burger, Tal Kerret, Lisa Silverstein, Roger Silverstein,
Larry Silverstein, and Richard Paul at the ribbon-cutting ceremony
for 3 World Trade Center on June 11, 2018

Then on a warm June Monday in 2018, we had the opening ceremony for 3 World Trade Center. I spoke that day, and my words were heartfelt. "Our first duty was to create a stirring memorial to those friends, neighbors and co-workers lost on 9/11," I told the many guests I had invited. "But it didn't stop there," I pointed out with a surge of emotion. "We were also charged with producing a more vibrant and connected neighborhood than anything that existed here before. A better version of New York."

A better version of New York. It was a vision I had kept firmly focused in my mind's eye since the grim morning of September 12, 2001. And finally seventeen long and arduous years later, I felt I had finally made it a reality. There were now three tall, inspiring skyscrapers as well as an evocative memorial and an inventively stylish transit hub standing on what had once been a charred ruin. I felt enormously gratified about what had risen up from the ashes. And I felt enormously proud of the role I had played in all that had been accomplished.

Yet I also knew there was still more that I needed to get done.

TOWER TWO WAS THE last piece in the puzzle. Back when the master plan had been revised in 2006, I had agreed to put up three towers on the original World Trade Center site. It had taken over a decade, yet I had succeeded in getting two of them built. But getting this final skyscraper completed, was, I had known, never going to be easy.

For one thing, it was a big, expensive building. Designed by Lord Norman Foster, the eighty-eight-story building with its slanted glass roof divided into four diamond shapes would be, at 1,349 feet, the tallest of my three towers. And it would be the most expensive. The construction costs were initially estimated at anywhere from $2.75 to $3 billion.

And for another, there would be no possibility of my once again using the Port's books to backstop the financing. Nor would I be able to use Liberty Bonds; all the available tax-free federal bond assistance provided for rebuilding after 9/11 had been allocated to the construction of my two other towers. I would need to raise the money—and $3 billion was certainly a significant amount—entirely on my own.

Which meant that before I could even think about going into the financial marketplace, I had to find a tenant. And not just any tenant.

I needed a blue-chip company that would want to lease at least half of the entire tower, about 1.5 million square feet of space.

But that was the thing I loved about New York—there would always be an employer out there who wanted his or her company to be in this city. Nowhere else could they find the vibrancy, the intellectual and creative excitement that fueled the "city that never sleeps." To paraphrase Sinatra, "If you can make it here, you can make it anywhere," and companies in all sorts of businesses, from the well-established old-guard firms to the newly burgeoning start-ups, had their hearts set on making it here. I just needed to find one of them.

AND I DID. Or so I thought. Citigroup, the global banking and financial services company, was looking for a new headquarters and told us they wanted to take the entire building. This was in 2013 and I was ecstatic. Here was the tenant I required in order to get my last building on the site erected. Once the lease was signed, I could use the rental income from Tower Two as a primary guarantee for the construction financing. It wouldn't be long before I would have all of my three towers up and running. I would have completed the last in a grueling series of challenges.

We spent a year and a half negotiating the lease with Citigroup. It was a rigorous process, but these things always are. I was satisfied by the final result, and so was the bank's executive committee. All that remained was for the papers to be signed.

But then both the bankers and I were blindsided: Citigroup's board of directors refused to approve the lease. They worried that the bank had barely managed to stay afloat in the rough seas of the 2008–10 financial crisis. It was too soon after that unsteady time to make such an expensive corporate relocation. The bank would need to demonstrate profitability over a sustained period before, the board grumbled, they could approve a move out of the present headquarters, however antiquated the facilities.

Just like that, I was back to square one. I was once again staring up at an eighty-eight-story challenge. And, as I knew all too well, I wasn't getting any younger.

BUT I ALSO BELIEVED people would want to come downtown. People would want to work in a lively, youthful neighborhood that was an innovative alternative to their grandfather's stodgy Midtown or Wall Street. And as it happened, this was precisely what first attracted James Murdoch, the chief executive officer of 21st Century Fox (and the younger son of News Corp's CEO, Rupert Murdoch), to the Tower Two site.

He decided the building would be the perfect location for the new headquarters of the Murdoch media companies, both 21st Century Fox and News Corp. In June 2015, the two media companies signed a letter of intent to occupy 1.5 million square feet at 2 World Trade Center.

But there was a catch. James Murdoch didn't like the tower that Foster had designed over a decade earlier. He thought it was more suited for a dull investment bank than an innovative, very much of the moment media company. And he brought in a brash young architect, Bjarke Ingels, to reimagine the skyscraper that would be built along Greenwich Street.

Ingels came up with something unique. It was a stepped tower, seven rectangular boxes piled on top of one another alongside set-back outdoor gardens. It looked like a twenty-first-century version of an ancient Mesopotamian ziggurat. Yet at the same time it shrewdly catered to the client's needs. There were capacious spaces for the Fox newsroom as well as television studios. A winding staircase set against the glass exterior wall connected the various companies in the Murdoch media empire, a way to promote the sort of corporate synergy the young Murdoch wanted. And, in a bit of Hollywood flamboyance, at the tower's summit, there was a Fox screening room that offered a breathtaking view over the city.

I could see why James liked the design, and I could appreciate how the building would work well for Fox. Yet it didn't work for me. I didn't like it.

But then I found myself looking in the mirror one morning while I was shaving. I stared at the face of an old guy; after all, I was eighty-four. At that moment I remember telling myself, "Silverstein, you got to be more flexible." And when I started thinking this way, I looked at the design again (and again, in truth; it took me some time) and I came to believe Ingels had conceived a great building. I really did. I found myself truly appreciating its uniqueness.

Still, I was concerned about how the Ingels design would fit in with the other towers on the site. So I went straight to the architects who had conceived them. I told these distinguished professionals I wanted their honest opinions. David Childs, Richard Rogers, Fumihiko Maki—they all were enthusiastic. They all agreed that the building reflected a more contemporary design language, yet one that would mesh well with the buildings they had created nearly ten years earlier.

But there was someone who didn't like it. And he was the one who needed to write the check. Rupert Murdoch telephoned one afternoon to ask, "You like the Ingels tower?"

Right away I knew I had a problem. But I nevertheless told him the truth. "I didn't at first, but I got to like it."

"Well, I don't," said Murdoch firmly.

So I went to work trying to persuade Rupert. I argued that we were both the same age, and we were just not used to new things. It took guys our age a while to come to terms with something new. How about, I suggested at last, you come here and I will arrange to have the architect make a presentation?

Murdoch agreed. And on the appointed day, Ingels, in his articulate, showman's way, was compelling. It was a wonderful presentation. Sure, James and his father were going at it a bit; the son certainly wasn't deferring to the company's CEO. But by the time Rupert left, he was in full accord. "I now see what you saw in the building, Larry," he told me.

Larry Silverstein and Bjarke Ingels discuss the new design for
2 World Trade Center in March 2015.

We went to work getting the documents ready to be signed. As always, the negotiations were tough. But as the months passed and we got to the morning of the final day when Murdoch could still exercise his contractual right to pull out, I was confident. This was a deal that would get done. I would get my final tower built.

Once again I was wrong.

At noon on that final day—November 11, 2016—I got a call from Rupert.

"Larry," he said straightaway, "I hate to tell you this, but this doesn't seem the best time for us to go forward."

"Rupert," I tried, "you already put thirty-five million dollars into this project."

"Larry, it's just that I feel that the economy is terribly fragile at this moment. It wouldn't be the best thing for my company to take on something like this."

We went back and forth a bit, but at last I had no choice but to say, "Thank you for your candor. That's life, I guess."

I made sure to end the conversation without expressing any hostility. You learn in business never to burn bridges. You want to keep relationships. You never know what tomorrow will bring.

But I was only putting up a good façade. In truth, I was devastated. We had put eighteen months into working with the Murdochs. I had been completely persuaded this was a done deal. It was a blow that left me deeply depressed.

Still, I wasn't ready to give up. If I have learned anything in all my years, it was that nothing in life was easy. To get something done, it took a lot of work. You needed to learn to put the defeats aside and move forward. I was still totally committed to getting Tower Two built.

AND IN 2022, I thought maybe I had found a way to get it done. Norman Foster had gone back to his drawing board and come up with a new, more contemporary rethinking of his original design. The new reiteration of Tower Two would be a 3-million-square-foot vertical campus of buildings within buildings that would rise 1,250 feet above Lower Manhattan; it was a vision that would work particularly well in the post-COVID era, when office life needed to be reimagined. And it would come with a new, very hefty price tag: $5 billion.

But I believed I had discovered how to get the bulk of the money I needed to build. Working hand in hand with the Port, Silverstein Properties had requested $3.9 billion from the federal government under the Railroad Rehabilitation and Improvement Financing fund. It was a very reasonable strategy. Tower Two would be directly connected to the PATH World Trade Center line as well as the MTA subway lines, and in addition critical mechanical, electrical, and plumbing elements that service the transportation infrastructure for the entire complex would be housed in the building's base. I was optimistic that the Transportation Department would see the logic and the necessity in giving us this funding.

As I said, you don't give up. You keep looking for new opportunities. And after more than two decades trying to get Tower Two built, I think I have found a path to getting it done. I am ninety-three years old, and the prospect is damn exciting. It makes me feel like a kid again. Or at least seventy.

AND ANOTHER OF MY hard-learned axioms was also paying new dividends. "Don't burn bridges"—that was how I conducted business for over seventy years. "You never know what tomorrow will bring." Well, despite all our problems and years of disputes, I am working again with the Port.

On the site of the demolished Deutsche Bank Building on Liberty Street at the southern tip of the Trade Center complex, in partnership with the Port and Brookfield Properties, I will be erecting Tower Five. It wasn't part of my original deal with the agency. Rather, it was an opportunity that only recently came my way, and I jumped at it.

And this time around, my dealings with the Port have been completely amicable. It has been a genuine collaboration. I give Rick Cotton, the executive director of the agency since 2017, all the credit for the way the Port had been transformed. He had brought a new sense of responsibility and professionalism to the agency. I never thought I would say this, but working with the Port on this new building has been a pleasure.

Tower Five will be a 900-foot skyscraper that will include affordable housing (four hundred of the units permanently rented at below-market value) as well as office, retail, and community space. It will be the first and only residential tower to be built on the sixteen-acre site.

And do you know why I decided in 2021 that an apartment tower needed to be built? Why it was, in fact, necessary to bring more housing downtown?

Because out of the ashes and ruin of 9/11, an energetic neighborhood where people want to live and work has been created. Because I have done what I'd set out to do. I have honored the victims. And I rebuilt the city I loved.

Foster and Partners design for the new 2 World Trade Center,
the last office tower at the World Trade Center site

LOOKING BACKWARD,
LOOKING FORWARD

ONE OF THE THINGS about being in your nineties is that you realize there is no point in trying to fool yourself. You have to face up to the facts. You have to look at what you have accomplished with a hard, unstinting gaze. There is no longer any reason not to come to terms with the life you have lived.

I go to the office nearly every day—or at least when Klara and I are not off boating on the *Silver Shalis*—and from my desk on the thirty-eighth floor of Tower Seven, I have a bird's-eye view of the entire sixteen-acre World Trade Center site. I look out through a wall of floor-to-ceiling glass at a neighborhood that didn't exist twenty-three years ago. A neighborhood that I helped to create.

I can still vividly recall those grim days after 9/11. I can remember the rubble, the toxic smell in the air, and the searing pain in the collective heart of an entire city. And I can still hear in my mind the naysayers, however sincere, who had lectured that no one would ever again want to work in buildings at the World Trade Center site. That no one would ever want to live in a grieving, shell-shocked downtown.

I told people that it would be a mistake to bet against New York. The city would come back—and it would be stronger than ever, I predicted.

I told people that if you build it, they would come. First-class office buildings would always find tenants in New York.

And I told people that we had a responsibility to rebuild—to those who lost their lives in the terrorist attacks, to their families, and to the country. It was vital to demonstrate to the world that America doesn't cower in the aftermath of enemy attacks. Rather, we rebuild, and this time bigger and better. We turn tragedy into triumph.

And we have done just that. From my vantage point in 7 World Trade Center, I can see the three other gleaming towers on the site, each climbing high into the sky, each bustling with the productive activity of tens of thousands of workers. There is David Childs's One World Trade Center, the tallest building in the Western Hemisphere; millions of visitors come each year to its observation deck to gaze out at this utterly marvelous gem of a city. There is Richard Rogers's muscular 3 World Trade Center, home to cutting-edge companies in technology, advertising, and branding. There is 4 World Trade Center, Fumihiko Maki's elegant vision that, despite its hefty seventy-two floors, seems to drift like a floating cloud into the sky. There is Santiago Calatrava's Oculus, its spindly, outstretched wings already an urban icon. And, not least, there is the 9/11 Memorial, Michael Arad's lyrically somber vision of two waterfalls cascading endlessly into dark pools, an attraction for millions of contemplative visitors each year. All this stands where there once was only dust, charred rubble, and the incinerated remains of those who tragically died that day.

While downtown, a part of the city that many had blithely written off, has become New York's hottest neighborhood. The residential population in the district has tripled since 9/11. It is a thriving community filled with young families, artists, and professionals, a neighborhood energized by schools, apartment complexes, stores, and restaurants. It is always hopping.

Only these days, the naysayers are back. There are people saying that because of the pandemic, New York will never be the same. That its offices will remain empty. Its restaurants will remain closed. And its residents will flee, escaping to the dubious safety of the suburbs.

All I can say to that kind of foolish, shortsighted thinking is that if the nine decades I have lived in this city have demonstrated one thing, it is this: Don't count New York out. This city always bounces back. People will always want to be here. People will always want to work here, live here, and play here. The day will inevitably come when the pandemic will become only a memory, and in its aftermath, New York will still be thriving, the capital of the world.

Yet I admit I had my days filled with doubts while rebuilding the Trade Center. I had spent years working hard, often against seemingly impossible odds. And I had suffered through my share of setbacks. There were times when I doubted whether I would be able to find the will to bounce back, to push forward. There were times when I wondered whether my good friend Lenny Boxer had done me a favor by so brilliantly engineering the legal strategy for the deal that had allowed me to buy the original Twin Towers. The reconstruction was a seemingly never-ending battle. There were so many powerful forces aligned against me. I found myself battling ambitious governors, wrongheaded mayors, incompetent bureaucrats, greedy insurance companies, and an often vindictive press. I genuinely feared that I would be overwhelmed by all these opponents. It was exhausting. The constant personal attacks were draining. I grew to feel that I would be forced to give up and walk away from my dream.

And I might have—if it had not been for Klara. It was her support that got me through all these battles. It was her love that helped me put all the negativity that was being hurled at me in perspective. It was her willingness to listen to my hours of complaints about the unfairness of all the attacks, the misguided personal criticisms, the willful ways of the politicians that gave me the strength to persevere. Her unwavering support, her wise guidance, her unqualified love—

these were absolutely essential. Without Klara to come home to each evening, I could not have gone off to do battle the next day.

With her support, I was able to continue my pursuit of my one inflexible goal: to help make New York the city it is, and the city it will be. I have devoted the last two decades of my life to bringing the World Trade Center back to life, to making sure that this city would honor its past and have a glorious future. And I look back at what I accomplished with a fierce and sustained pride.

But now that I'm in my nineties, the responsibilities of age require me to make a full and candid accounting of my life. And while rebuilding the Trade Center site was indeed one of my most cherished, prideful accomplishments, it was not, I confess, my greatest.

I have built something that is more impressive, more valuable than any skyscraper, no matter how tall it stands. I have built a family. I have been blessed to be married to Klara for more than sixty-eight years. I have three children, eight grandchildren, and one darling great-grandchild. My son, Roger, was in charge of leasing our properties for many years; it was his skill and tenacity that helped make Tower Seven such a success when the naysayers were warning that we would never rent the space at the prices we were asking. He is now semi-retired, but still an active advisor when the big decisions need to be made. Sharon, my eldest child, went off to Stanford, worked with me at Silverstein Properties for a few years, then got an MBA from Harvard Business School, and wound up returning to the warm West Coast weather to be a full-time mother. My youngest, Lisa, is now chief executive officer and vice chairman of Silverstein Properties. At this point, I have handed the baton to Lisa and she is the one really making all the day-to-day decisions. Sure, she is assisted by her gifted husband, Tal, the company president, and a terrific team of executives and employees. But it is my daughter Lisa whom I have appointed to lead the company into the future. And so while we have a portfolio that includes more than 15 million square feet of property and over $10 billion worth of properties in development, it is still very much a family business. Just as it was when I first started out seventy years ago with my dad.

It is this family that is my indelible legacy. This is my true accomplishment. It is what has sustained me through all the many battles, and it is what continues to give me immeasurable pride as I move full of wonder and great expectations into all that is still to come.

ACKNOWLEDGMENTS

Rebuilding the World Trade Center was a collective effort involving thousands of people. Many of them put their hearts and souls into this project. It was not easy, and there were many disagreements and setbacks. We were driven by our unshakable love for New York and by the need to respond in a way that would make everyone proud.

Our goal was to commemorate those we lost and to create a place full of life and creativity that would reflect who we are as New Yorkers. Today the new World Trade Center has come alive as a dynamic public space with timeless architecture. It is home to some of the city's most exciting companies.

I couldn't have done any of this without my family, colleagues, friends, neighbors, tenants, and my partners in government.

I would like to begin by acknowledging and thanking my wonderful wife of sixty-eight years, Klara. I certainly wouldn't be here today had it not been for her boundless support, wisdom, and encouragement. She was right by my side on September 11, 2001, and has been my partner ever since. She encouraged me to commit the rest of my life to rebuilding the World Trade Center, and she vowed to do the same.

ACKNOWLEDGMENTS

I've often said the most important decision you make in your life is to choose the right spouse or partner. The wrong one is a disaster, but the right one enables you to do anything you want to do with your life, because you do it together. You do it as partners. You're totally dedicated to each other. You're there for each other without any hesitation or equivocation, without any reservation. It's total, it's a commitment. Thank you, Klara, for being my partner—in everything.

I have said before that my family is more important than any skyscraper. To my three children, Sharon, Roger, and Lisa; my son-in-law, Tal; my daughter-in-law, Patricia; my eight grandchildren: Ariel, Cory, Eli, Joshua, Julian, Rachel, Robert, and Zachary; and my one darling great-grandchild, Harry—thank you for your unconditional love, encouragement, and support.

My parents, Etta and Harry, gave me my values, for which I am eternally thankful. I am grateful to have passed on some of those values to my children and grandchildren.

I am delighted that after working side by side for over thirty years, my daughter Lisa has taken over the leadership of our company, Silverstein Properties. We think alike and share many of the same values. I have no doubt that this company will be in great hands under her skillful guidance.

Writing this book has been a wonderful experience over many years. A very big thanks to Howard Blum for being my guide, sounding board, and partner along the way. Thank you to Ellis Levine and Eric Lupfer for your wise and strategic counsel. My fantastic publishing team at Knopf, who were so passionate about this project from the very beginning: Reagan Arthur, Quynh Do, Andrew Miller, Anne Achenbaum, Sara Eagle, Erinn Hartman, Laura Keefe DeLange, Micah Kelsey, Chip Kidd, Judy Jacoby, Loriel Olivier, Nicole Pedersen, Emily Reardon, Tiara Sharma, and all the wonderful folks at Alfred A. Knopf—thank you for your great work over the years.

I wouldn't have purchased the World Trade Center in 2001 without my partners Lloyd Goldman, Jane Goldman, Joe Cayre, and Frank Lowy. Thank you for remaining steadfast in the commitment to rebuild, against all odds.

I would also like to thank my dear friend Bill Berkley for his support, advice, and encouragement after 9/11.

Of course, I couldn't have rebuilt the World Trade Center without my team at Silverstein Properties. Janno Lieber, thank you for your leadership of this historic and extraordinarily challenging project from the very beginning. Mickey Kupperman, thank you for coming out of retirement twenty years ago to help me with the difficult task of managing the design of three World Trade Center towers on a ridiculously tight deadline.

David Worsley started working on the World Trade Center project twenty years ago, and has expertly guided its design and construction ever since, along with a team of some of the most talented and hardworking professionals in the business: Fred Alvarez, Michiko Ashida, Lai Mei Chau, Dan DeLosa, Serge Demerjian, Ken DeRenzis, Mary Dietz, Duan He, Jon Hong, Sean Johnson, Nate Matisoff, Molly Mikuljan, Stefanie Moreo, Ed Narbutas, Nygel Obama, Guy Punzi, Scott Thompson, Carlos Valverde, and Malcolm Williams. Jack Klein was responsible for building 7 World Trade Center, the first tower to rise at the World Trade Center site. On the business side, Mike Levy and Marty Burger led the successful financing of 3 World Trade Center.

I would like to thank my leadership team, Lisa, Tal, Mickey, and Dino Fusco, as well as current and former members of our terrific management team for the wonderful work they do managing their own teams, day in and day out: Lisa Bevacqua, Chi Chu, Bill Dacunto, Jeffrey Deitrich, Debra Hudnell, Jon Knipe, Katie Kurtz, Michael Levy, David Marks, Lior Mor, Jeremy Moss, Nick Pazich, Leor Siri, Yael Urman, Guy Vardi—thank you from the bottom of my heart.

There are many other amazing people at Silverstein Properties whom I consider my second family. I cannot name you all, but I would like to acknowledge my wonderful executive assistant, Shoshana Hill. I will never forget her predecessor, Ann Tobin-McKevitt, who worked by my side for thirty-seven years. Thank you also to Debra Hudnell, Pam Brown, Janet Clarke, and Pedro DeVargas, who have also been by my side as far back as I can remember.

I would like to thank Dara McQuillan, the company's head of marketing and communications, for helping me tell the story of the new World Trade Center through books, films, television shows, exhibits, and art. Over the past two decades, Dara has worked with a team of some of the most talented communications professionals in the city, including Colette Bosque, Anna Colavecchio, Caroline Grassi, Aisling Gregory, Gianna Frederique, Jessica Schoenholtz, Rebecca Shalomoff, and Courtney Squicciarini Newton.

It has given me great pleasure over the past twenty-three years to work with some of the most talented and successful architects and designers in the world. I am grateful for their mentorship, vision, and commitment. Thank you in particular to David Childs for your extraordinary work on 7 World Trade Center and One World Trade Center. An enormous thanks also to Daniel Libeskind for creating an inspiring master plan for the new World Trade Center. You set the stage for everything we have done here. Thank you to the architects who brought Daniel's plan to life: Michael Arad and Peter Walker, Santiago Calatrava, Norman Foster, Fumihiko Maki, Joshua Ramus, Richard Rogers, and the team at Snøhetta, and also Bjarke Ingels for your creative work on 2 World Trade Center. I can't think of anywhere else on earth where such a talented group of architects worked together to design the individual components of a master plan on such a scale.

Of course, each architect has a team of hardworking and creative architects and engineers working alongside them. Much appreciation goes to Björn Andersson, Matthew Burton, Chris Cooper, Augustine DiGiacomo, Bob Ducibella, Scott Frank, Carl Galioto, Richard Garlock, T. J. Gottesdiener, Peter Han, Mike Jelliffe, Gary Kamemoto, Shawn Kirkham, Joyce Lam, Ken Lewis, Shig Ogyu, Richard Paul, Tosan Popo, Ahmad Rahimian, Osamu Sassa, Jeffrey Smilow, Brian Towers, Craig Tracy, Michael Wurzel, Nick Zigomanis, and hundreds of other hardworking professionals.

As much as it was a herculean task building these new office towers, it was just as difficult leasing them. We couldn't have done it without our amazing leasing team led by Roger Silverstein, Jeremy Moss, Mary Ann Tighe, and her colleagues at CBRE. Thanks to them, the new World Trade Center is now home to hundreds of the city's most interesting and successful companies from Moody's to WilmerHale, Spotify to WPP and GroupM, McKinsey, Diageo, Moët Hennessey, and many, many others. They wouldn't have moved here without the guidance and hard work of James Ackerson, David Caperna, Steve Eynon, Howard Fiddle, Adam Foster, Brad Gerla, Christie Harle, Evan Haskell, Robert Hill, Caroline Merck, Ken Meyerson, Peter Turchin, and Simon Wasserberger. I would also like to take a moment to thank my old friend and colleague Steve Siegel who worked with me and the team to lease 7 World Trade Center, and Keith Cody and Joe Artusa who are doing a terrific job today.

Thank you also to the many talented tenant brokers for showing your clients that the World Trade Center is one of the best places in New York to grow a business. I am grateful for your work on behalf of the companies that have made the World Trade Center their new home: Matthew Barlow, Michael Burgio, Joseph Cabrera, John Cefaly, George Chatzopoulos, Alexander Chudnoff, Andrew Coe, Robert Constable, Christopher Corrinet, Lauren Crowley Corrinet, Gary Davies, Lloyd Desatnick, Eric Deutsch, Douglas Dolgoff, Frank Doyle, Sam Einhorn, Robert Eisenberg, Alice Fair, Jonathan Fein, Lee Feld, Augustus Field IV, Jeffrey Fischer, Jared Freede, Michael Geoghegan, David Glassman, Neil Goldmacher, David Goldstein, Robert Goodman, Jason Gorman, Michael Gottlieb, Scott Gottlieb, Gary Greenspan, Joseph Harkins, Patrick Heeg, Brendan Herlihy, Christopher Hogan, Martin Horner, Adele Huang, William Iacovelli, Jared Isaacson, Scott Klau, David Kleinhandler, Rocco Laginestra, Liz Lash, Sinclair Li, Michael Liss, Matthew Lorberbaum, Robert Lowe, Michael Mathias, Matthew McBride, Ryan McKinney, Ken Meyerson, Arthur Mirante, Paul Myers, Paul Nugent, Jennifer Ogden, Andrew Peretz, Marcus Rayner, Peter Riguardi, Michael Rizzo, Robert Romano, David Rosenbloom, Drew Saunders, Erik Schmall, Sam Seiler, Kenneth Siegel, Matthew Siegel, Mitchell Steir, Barrett Stern, Jarod Stern, Harley Stevens, Moshe Sukenik, Gregory Taubin, Mary Ann Tighe, Gregory Tosko, Dan Turkewitz, Munish Viralam, Zachary Weil, Mark Weiss, Bernard Weitzman, James Wenk, John Wheeler, Daniel Wilpon, Robert Wizenberg, and Sacha Zarba.

The immense progress that we've made in Lower Manhattan is a testament to this community's boundless faith and willingness to bet on New York. In the days, weeks, and months after 9/11, Senator Chuck Schumer was New York City's leading advocate in Washington. Together with Senator

Hillary Clinton, they secured the federal money that was vital to the recovery and subsequent rebuilding.

Although we didn't always see eye to eye on the details, many leaders of New York State shared my vision for a rebuilt World Trade Center. Thank you to Governor George Pataki, John Cahill and his team, Governor Eliot Spitzer and Eric Dinallo and their teams, Governor David Paterson, Governor Andrew Cuomo, and most recently Governor Kathy Hochul—thank you for your leadership, and for your focus on much-needed affordable housing at 5 World Trade Center.

Thank you also to Mayor Eric Adams and his administration, as well as to New York City Council Speaker Adrienne Adams and her colleagues on the City Council. I would also like to thank and acknowledge Senator Kirsten Gillibrand and Representatives Carolyn Maloney and Jerrold Nadler.

Of course, I owe an enormous debt of gratitude to Mayor Mike Bloomberg. Without Mike, we would not have opened the 9/11 Memorial on the tenth anniversary, the museum would be a very different place, and there would be no performing arts center. He has been a steadfast champion of the new World Trade Center as a place to remember, renew, and rebuild— and as a place of life, hope, and creativity. Early on, Mike had the support of Dan Doctoroff, Patricia Harris, NYPD Commissioner Ray Kelly, Kate Levin, Bob Lieber, Seth Pinsky, and many other talented professionals who committed their time to this project.

Mike's team was supported by the Lower Manhattan Development Corporation, where I had the pleasure of working with the late John Whitehead and Bob Douglass, as well as David Emil, Holly Leicht, and Avi Schick.

Most recently, it has been a pleasure working with everyone at the National September 11 Memorial and Museum, including Joe Daniels, Alice Greenwald, and Elizabeth Hillman.

I am grateful to have been involved over many years with the Perelman Performing Arts Center at the World Trade Center. That building opened last September and is a resounding success thanks to Mike Bloomberg, Ron Perelman, Maggie Boepple, Leslie Koch, Khady Kamara Nunez, Cathy Blaney, Bob Pilon, and many others.

The Port Authority has been my partner at the World Trade Center since 1980, when I won the bid to build the original Seven World Trade Center. I have seen many chairmen and executive directors come and go over the years, but I have never had as good a relationship as I have today under the leadership of Rick Cotton and Kevin O'Toole. Thank you for your support, cooperation, and creativity in making this place a success. I would also like to thank all the hardworking leaders at the Port Authority over the years, including the late Neil Levin, Tony Coscia, Pat Foye, Glenn Guzi, Steve Plate, Scott Rechler, Alan Reiss, Anthony Shorris, Derek Utter, and Chris Ward.

None of our new buildings here would function for one minute without our talented property management teams, engineers, cleaners, and security staff. Thank you all for the important work you do every day, often behind the scenes. In particular I would like to acknowledge our longtime head of operations Bill Dacunto, and his property management team: our current and former World Trade Center property managers Jim Halpin, Angelo Provvido, Alex Riveira, and Brian Smith; assistant property managers Amire Papraniku, Dayna Scalogna, and Yvette Wright; and chief building engineers Steve Nathan, Jay Siller, and John Ferrandino. Some of these people survived the 9/11 attacks and, like me, devoted the rest of their working lives to rebuilding and operating the new World Trade Center.

My company lost four employees at the World Trade Center on 9/11: James Corrigan, our fire safety director; John Griffin, our director of operations; Charles McGee, our chief engineer; and John O'Neill, our director of security. We will never forget you.

As the developer of the commercial part of the new World Trade Center, I had a special responsibility to ensure that everyone who wanted to participate in the project had an opportunity to do so. Working with organizations such as Nontraditional Employment for Women and programs such as Construction Skills 2000 and Helmets to Hardhats, we used this project to provide opportunities for tens of thousands of New Yorkers. Union construction jobs are one of the best ways to elevate New Yorkers to the middle class, and diversifying the building trades was one of our top priorities. I would like to thank everyone in the construction industry for their work at the World Trade Center over the past twenty years. Gary LaBarbera, Lou Coletti, the late Ed Malloy, and many other labor leaders, including 32BJ leaders Mike Fishman, Héctor Figueroa, and Kyle Bragg; Local 94 leaders Kuba Brown and Ray Macco; and Realty Advisory Board president Howard Rothschild. Enormous thanks to the Tishman team that built 7 World Trade Center, 4 World Trade Center, and 3 World Trade Center: Dan Tishman, Jay Badame, Mike Mennella, John Kovacs, Tom Leo, Dean Essen, Mike Pinelli, Bill Stanton, Mike Goldberg, Ted Carpinelli, and my friend Frank Hussey, who came down to Ground Zero on September 12, 2001, and spent almost twenty years working on each of our new buildings there.

Of course, nothing happens in business without the counsel of the legal profession. I am blessed to have had the support and advice of some of the greatest legal minds in the world, including my high school and college classmate Herb Wachtell and his partners Marty Lipton, Bernie Nussbaum, Marc Wolinsky, Robin Panovka, Adam Emmerich, Peter Hein, Eric Roth, Jonathan Moses, Ben Germana, and their many colleagues, who fought beside me for years to recover the insurance proceeds we needed for rebuilding, and who advised my team and me during the many negotiations with the Port Authority and other stakeholders. I would also like to thank Ben Needell and Marco Caffuzzi and their colleagues at Skadden Arps, Bruce Ficken, Karen Scanna, Michael Subak, and, of course, my old friend and counsel Leonard Boxer who worked around the clock to support my bid to buy the World Trade Center in July 2001.

Throughout this long and complicated public-private partnership, I have been fortunate to work with a team of wonderful political, government, and public relations advisors. I would like to extend a big thank-you to Jon Silvan, Justin Lapatine, and the team at Global Strategy Group; Roberto Ramirez, Luis Miranda, Eduardo Castell, and Catherine Torres at the MirRam Group; Suri Kasirer, Julie Greenberg, Eldad Gothelf, and Melissa Rosenberg at Kasirer; my old friend the late Howard Rubenstein and his son Steven, as well as Bud Perrone, Suzi Halpin, Nancy Haberman, Justina Lombardo, and Gerald McKelvey at Rubenstein Communications; Mike McKeon, Mike DuHaime, and Roger Bodman; Jonathan Rosen, Sara Joseph, Jeremy Soffin, and the entire BerlinRosen team; my early advisors—

the late Bill Plunkett, Peter Powers, David Samson, and Brad Card; and Jack Quinn and Ed Gillespie for being my guide and introduction to key members of Congress in the months after 9/11.

The media played an important—and often challenging—role in this project. Of course, it is a reporter's job to ask difficult questions, and to relay what they see and hear. I would like to express my gratitude to the reporters, editors, and news producers who covered this historic project over many years, and in particular the ones who never gave up on New York, downtown Manhattan, and me: Steve Cuozzo and Lois Weiss, longtime real estate reporters for *The New York Post,* and always big champions of a rebuilt World Trade Center; Scott Raab of *Esquire* magazine for dedicating ten years of his life to covering the project; Dan Geiger and Max Gross and everyone at the *Commercial Observer* for their expert coverage of the real estate industry; Dave Stanke for his unwavering support in the difficult early years of the project; Matt Kapp for his insight into what makes Lower Manhattan such a diverse and fascinating neighborhood; Grace Capobianco, who created *Downtown* magazine as a result of what happened on 9/11, and who has diligently chronicled the comings and goings of this neighborhood ever since; and Scott Pelley and his colleagues at *60 Minutes* for shining a light on the delays and frustrations at the site, which inspired our political leadership to hold us all accountable.

While we were rebuilding the World Trade Center, downtown has emerged as a new model of what is best and most dynamic about New York. The population has tripled, and the residential neighborhood around the World Trade Center is now one of the city's most desirable places to live and raise a family. The leadership of our downtown community showed great perseverance, patience, and encouragement as we rebuilt. Catherine McVay Hughes, Julie Menin, Jessica Lappin, Andy Breslau, and the late Elizabeth Berger—it was a pleasure working with you, as well as members of the Manhattan Community Board 1, and everyone in the neighborhood Klara and I call home.

Art has always played a big role in my personal and professional life. When we rebuilt 7 World Trade Center, I insisted that we bring in artists at the earliest possible stage of the design process, including Jenny Holzer,

ACKNOWLEDGMENTS

Jamie Carpenter, Ken Smith, and Jeff Koons. As I set out to rebuild the rest of the World Trade Center, I felt it was important to document the massive rebuilding project through photography, film, time-lapse photography, and art. Our photographer Joe Woolhead has shot over four million photographs of the rebuilding effort since 9/11, including many featured in this book. Mike Marcucci has been filming for us since 2004, and in 2012, he produced the award-winning documentary, *16 Acres.* When we opened 7 World Trade Center, a handful of fine artists worked in the building, including Diana Horowitz, Marcus Robinson, Todd Stone, and Jacqueline Gourevitch. Over the years they were joined by many more, including Kerry Irvine, BoogieREZ, Stickymonger, Lady Aiko, Clayton and Parker Calvert, and many other incredibly talented people—each of them bringing life and creativity to the new World Trade Center. We recently installed a magnificent sculpture, *Jasper's Split Star,* by the late Frank Stella in the Silverstein Family Park outside 7 World Trade Center.

A few years ago, my grandson Cory launched Silver Arts Projects, a nonprofit organization dedicated to supporting emerging artists. The residency is run by Gregory Thornbury with the support of Agnes Gund, Jared Owens, and Lilly Robicsek. Thank you also to Eric Widing for your support launching this wonderful initiative.

Klara and I are immensely grateful for all the wonderful people we've met over the course of our lives and the ensuing friendships that continue to mean the world to us. We are both grateful to know all of you, and we look forward to many more years of friendship to come.

In conclusion, I'd like to thank you all for being such an important part of our lives, and the life of this city we call home. I have always been bullish on the future of Lower Manhattan and New York City. But I've been far from alone on that. Our experience rebuilding the World Trade Center proves that when passionate New Yorkers work together, we can overcome anything and achieve anything. I think I can speak for all of us when I say how proud I am to have been a part of this historic endeavor. Together we have reinvented what it means to be a city for the twenty-first century.

Larry A. Silverstein

INDEX

Page numbers in *italics* refer to illustrations.

ILLUSTRATION CREDITS

ILLUSTRATION CREDITS

COLOR INSERT